Cyclodextrins - Core Concepts and New Frontiers

Edited by Rashid Ali

Published in London, United Kingdom

Cyclodextrins - Core Concepts and New Frontiers
http://dx.doi.org/10.5772/intechopen.102262
Edited by Rashid Ali

Contributors
Syed Haroon Khalid, Rimsha Yousaf, Fizza Abdul Razzaq, Sajid Asghar, Muhammad Irfan, Ikram Ullah Khan, Ludmila Aricov, Gabriela Ioniță, Anca Ruxandra Leonties, Iulia Matei, Olga Fedorova, Yuri Fedorov, Rashid Ali, Ishfaq Ahmad Rather, Ahmad Hasan

First published in London, United Kingdom, 2023 by IntechOpen
IntechOpen is the global imprint of INTECHOPEN LIMITED, registered in England and Wales, registration number: 11086078, 5 Princes Gate Court, London, SW7 2QJ, United Kingdom

British Library Cataloguing-in-Publication Data
A catalogue record for this book is available from the British Library

Additional hard and PDF copies can be obtained from orders@intechopen.com

Cyclodextrins - Core Concepts and New Frontiers
Edited by Rashid Ali
p. cm.
Print ISBN 978-1-80356-257-5
Online ISBN 978-1-80356-258-2
eBook (PDF) ISBN 978-1-80356-259-9

Meet the editor

Dr. Rashid Ali is an assistant professor in the Department of Chemistry, Jamia Millia Islamia, New Delhi, India, and a pioneering researcher engaged in supramolecular and organic chemistry. He obtained his Ph.D. from the Indian Institute of Technology (IIT) in Bombay, India. He has more than 12.5 years of research experience, one of which he spent at Sookmyung Women's University, Korea. He has published more than sixty original research papers in scientific journals as well as several book chapters and one book. Dr. Ali is a reviewer for numerous international journals.

Contents

Preface

Since Villiers first described "cellulosine" in 1891 and Schardinger laid out the fundamentals of cyclodextrins (also known as cycloamyloses) in the early nineteenth century, the chemistry of cyclodextrin (CD) and its congeners has continuously attracted interest from the scientific community worldwide. This valuable scaffold has numerous potential applications, including in solar cells, semiconductors, super-capacitors, polymers, drug delivery, sensors, ligands, nanotechnology, biomedicine, agriculture, dyes, food technology, cosmetics excipients, textiles, and the pharmaceutical industry, in addition to its role in the environmental, biological, supramolecular, and analytical horizons. The most distinctive signature of CDs is their capability to form inclusion complexes with a variety of small molecules through host–guest supramolecular interactions.

This book deals with core concepts and new perspectives on CDs. Chapter 1 describes the historical background of CDs along with their physiochemical characteristics, as well as provides a short overview of their diverse applications. Chapter 2 discusses the fundamental concepts of CDs, including their structure, properties, and host–guest interactions with a particular emphasis on molecular dynamics in addition to drug delivery uses. Chapter 3 highlights recent advancements in sensing diverse biologically important small molecules by CDs through colorimetric, fluorescence, electrochemical, and potentiometric responses. Chapter 4 describes the inclusion chemistry of various organic dyes with CDs. Finally, Chapter 5 focuses on CD-based polymers (both natural and synthetic) in solution as well as in gel states.

I am grateful to all the chapter authors for their excellent contributions. I also express my sincere thanks to my wife Saba Khan and my lovely daughters Naira Khan and Samaira Khan. Finally, I am also thankful to my family members, relatives, colleagues and friends for their continuous motivation and encouragement without which this book would not have been possible.

Rashid Ali
Department of Chemistry,
Faculty of Natural Sciences,
Jamia Millia Islamia,
New Delhi, India

Chapter 1

Introductory Chapter: Historical Background – Fundamental Structural and Physiochemical Properties of Cyclodextrins (CDs)

Rashid Ali

1. Introduction

The discovery, by Antoine Villiers, of the biosynthetic cyclic oligosaccharides-based seminatural products consisting of 6, 7, and 8 chiral glucose units, arranged in a donut shape and connected *via* α-1,4-gycosidic bonds, is generally symbolized as α-, β-, and γ-cyclodextrins (or ACDs, BCDs, and GCDs), respectively. They seem to be the most investigated macrocyclic host molecules in the realm of supramolecular chemistry—a study of the noncovalent interactions. Naturally, they are being obtained from the enzymatic degradation of one of the most indispensable polysaccharides, that is, potato starch in the bacteria (**Figure 1**) [1]. Sometimes, they also dubbed as the enzyme-modified starch derivatives. These macrocyclic systems comprise the lipophilic inner cavities as well as hydrophilic outer surfaces of the particular interest. Interestingly, the cyclodextrins (CDs) are produced "hundreds-of-thousands" of tons every year by means of environmentally friendlier, simple yet effective techniques and methods. The CDs belong to a family of "cage molecules" in which the core of their structures is unruffled of a very stable hydrophobic cavity, having the distinctive property of encapsulating the hydrophobic entities by virtue of the invaluable host-guest supramolecular interactions. The driving strengths that operate in the inclusion complex formation are van der Waals and electrostatic interactions besides the hydrogen bonding forces. Generally, the complex formation by the CDs depends on the shape and size of the cavities of CDs, chemical nature of the guests, expulsion of the high-energy H_2O molecules, and CD-CD aggregation. More importantly, their vibrant properties can easily be altered significantly through their ability of forming the inclusion complexes and also by means of their apposite functionalizations, as they contain a groups of primary as well as secondary hydroxyl functionalities at the two rims (**Figure 2**) [2]. The chemical structures of the most popular cyclodextrins, i.e. α-, β-, and γ-CDs, are depicted in **Figure 3**. As shown in **Figure 3**, the CDs have "truncated cone shape" rather than the perfect cylindrical structures because of the chair conformations of the glucopyranose units present in these types of cyclic systems. The toroidal structure of the CDs contains a panel of secondary hydroxyl groups on the wider rim, whereas the primary hydroxyl groups are present at the narrower rim side. The hydrophobic cavity is clearly displayed with an arrow inside the truncated cone as displayed in **Figure 3**. On the other hand, different structural

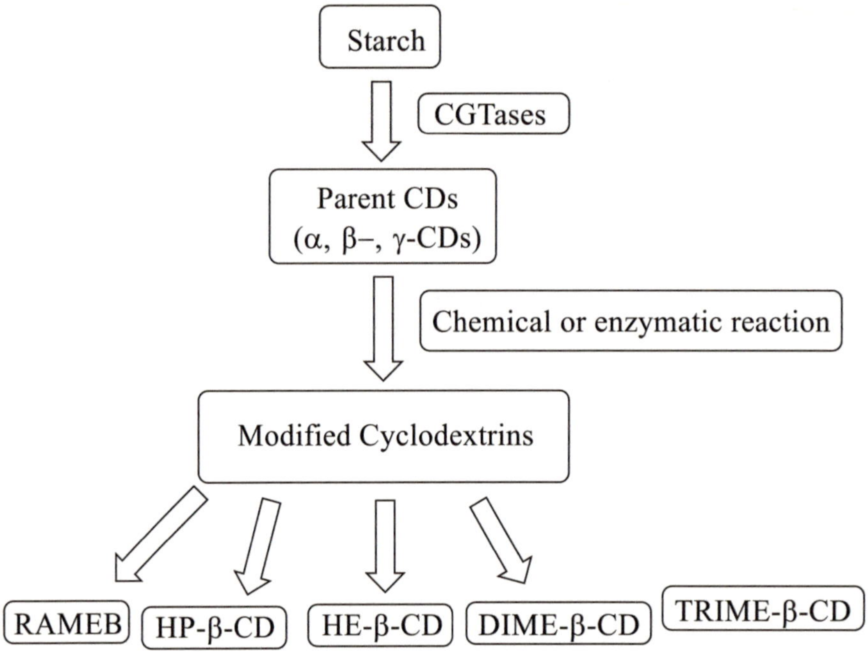

Figure 1.
Parent as well as the modified CDs of the pharmaceutical importance: [RAMEB: randomly methylated β-CD; HP-β-CD: hydroxy propyl β-CD; HE-β-CD: hydroxy ethyl β-CD; DIME-β-CD: heptakis-2,6-dimethyl-β-CD; TRIME-β-CD: heptakis-2,3,6-trimethyl)-β-CD].

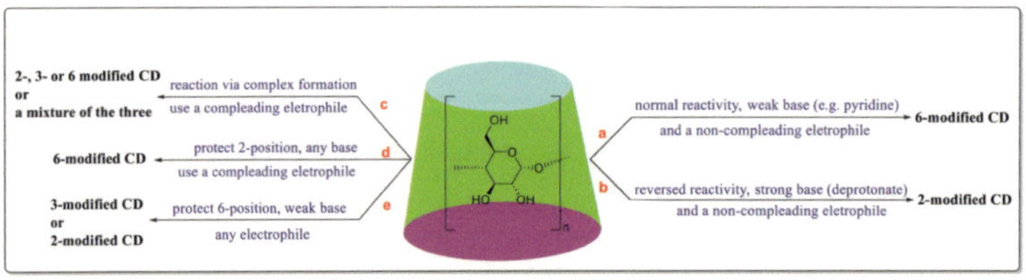

Figure 2.
Various possible functionalization sites to generate a variety of modified CDs.

and physiochemical properties of the cyclodextrins are tabulated in **Table 1**—just to provide a quick glance to the readers so as to compare and have adequate knowledge of about these versatile parameters.

Remarkably, their unique "molecular encapsulation" signatures had already been immensely exploited in a myriad of industrial products, technologies, and analytical services as well. The fascination toward the researchers and industrialists worldwide could be inspected from their diverse potential applications in pharmacy, dyeing, food, medicine, biology, biomedicine, biotechnology, beverage industry, organic solar cells (OSCs), nanotechnology, environmental protection, wastewater treatment, conducting polymeric materials, semiconductors, supercapacitors, agrochemistry, remediation, "cosmetology and hygiene," catalysis, drug carriers, and ligands engineering, besides their usage in the chiral chromatographic separations (**Figure 4**) [3]. Moreover, the CDs had also been used as the crucial "bricks" in assembling the vital supramolecular architectures of the meticulous importance, such as catenanes,

DOI: http://dx.doi.org/10.5772/intechopen.107434

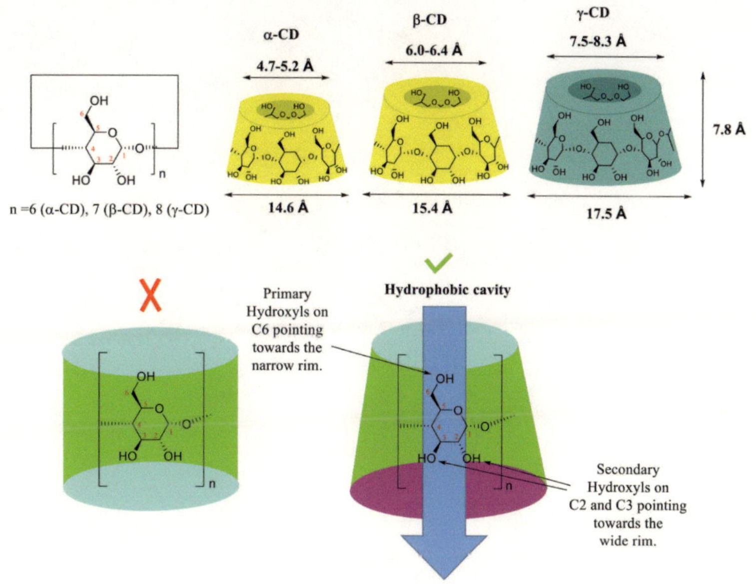

Figure 3.
Structures of α/β/γ-CDs (top), and wrong and correct pictures of CDs (bottom).

S. no.	Characteristics and properties	α-CD	β-CD	γ-CD
1	Number of glucopyranose units	6	7	8
2	Molecular weight (g/mol)	972	1135	1297
3	Internal diameter (Å)	4.7–52	6.0–6.4	7.5–8.3
4	External diameter (Å)	14.6	15.4	17.5
5	Height of torus (Å)	7.8	7.8	7.8
6	Volume of the cavity ($Å^3$)	174	262	427
7	Solubility in water at 25°C (%, w/w)	145	18.5	233
8	Partial molar volumes in solution (mL mol^{-1})	611	703.8	801.2
9	Crystal form (from water)	Hexagonal plates	Monoclinic parallelograms	Quadratic prisms
10	Diffusion constant at 40°C	3.443	3.224	3.000
11	pK (by potentiometry) at 25°C	12,332	12,202	12,081
12	Surface tension (mN/m)	—	71	71
13	50% hemolysis (mM)	—	5.3	8.5
14	Hydrolysis by *A. oryzae* α-amylase	Negligible	Slow	Fast
15	Melting temperature limits (°C)	255–265	255–260	240–245
16	Water content of the crystal (wt. %)	13.5–14.5	10.2	8.13-17.7
17	Cavity diameter inner (nm)	0.47–0.53	0.60–0.65	0.75–0.83
18	Cavity diameter outer (nm)	1.46	1.54	1.75

Table 1.
Different structural as well as physiochemical properties of the cyclodextrins.

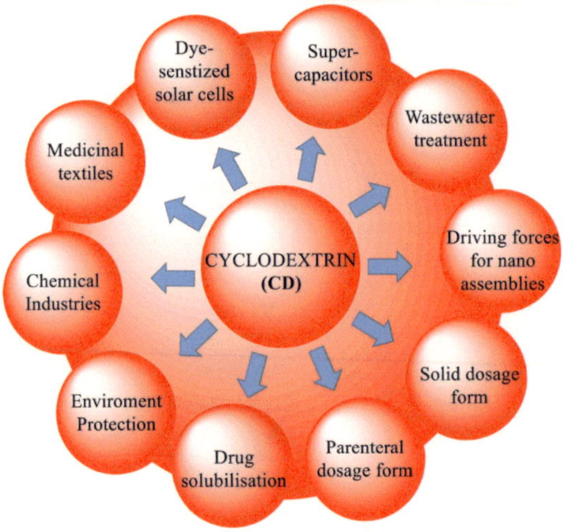

Figure 4.
Pictorial representation of diverse potential applications of the CDs.

rotaxanes, polyrotaxanes, supramolecular polymeric assemblies, and so forth [4]. Commercial products entailing the CDs, used in our daily lives, are displayed in **Figure 5** [5]. In this particular book based on the CDs, we intended to showcase the new frontiers in this emerging arena with an intention to aware the readers where this wonderful field presently stands, and where it might go in years to come, though fully matured. We anticipated that this new package in the form of book based on the CDs chemistry would be much informative to the researchers working in both academia and industry. Surely, it will also be very helpful to the undergraduate and postgraduate students in addition to the young minds planning to enter into the ever-booming area of research.

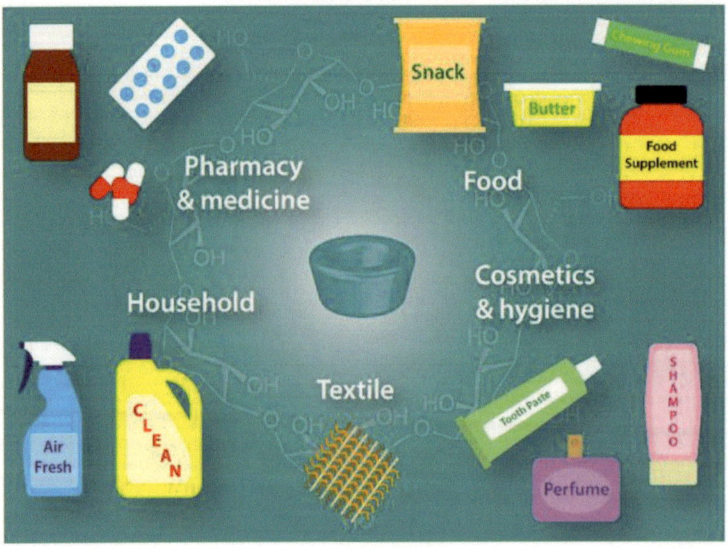

Figure 5.
Different commercial products consisting of the CDs in our daily-based lives.

2. Historical backgrounds of CDs

Noticeably, until mid-1970s, α-, β-, and γ-CDs were accessible only in small amounts, and they were contemplated as only the "laboratory curiosities." Because of their presumed toxicity and their high prices in addition to the unavailability of adequate knowledge, their industrial potentials were totally masked at that time. Although CDs have been well known for more than 130 years, they only truly "took off" in 1980s when for the first time "applications of the CDs" in pharmaceutical and food industries were successfully revealed. This progress was made by the production of the α-, β-, and γ-CDs on an industrial scale, and these systems were fruitfully achieved in extremely pure form in 1984. Freudenberg's research team in 1936 proposed the cyclic structures for both α- and β-dextrins, and in 1953, his group had published the first ever patent in this field related to the pharmaceutical formulations [6]. Remarkably, the low cost of these cyclic polysaccharides vastly impacted their long range developments, particularly that of the β-CD.

Nomenclature: During the groundbreaking discovery of the CDs in 1891 by A. Villiers (1854–1932), a France chemist and pharmacist, the CDs were dubbed as "cellulosine" because of their similar properties (i.e. nonreducing, crystalline, resistant to the acid hydrolysis, and water-soluble) as cellulose. Soon after these findings, F. Schardinger, the so-called "*Founding Father of the CDs*," identified the naturally occurring α- and β-CDs, and at that time denoted them as 'Schardinger sugars' but later on famed as cyclodextrins. Whereas, French in 1942 recommended Schardinger's dextrins as the cycloamyloses. On the other hand, the γ-CD was discovered by the research team of Schardinger in the year 1948. Moreover, Cramer in 1956 introduced and described the notion for inclusion complexes. The metal complexes involving CDs were achieved by means of the monotosylation approach in 1990s [7].

Timeline history: Noticeably, in a beautiful review published by G. Crini [1], history of the CDs was divided into *five* vital periods: (1) discovery by Villiers and the characterization chemistry by Schardinger in the period of 1891–1911; (2) the 25 years period (1911–1935) of doubt and disagreements (between Pringsheim and Karrer), but at that time Pringsheim from Germany was the leading researcher in this particular arena and demonstrated that CDs formed stable aq. complexes with numerous

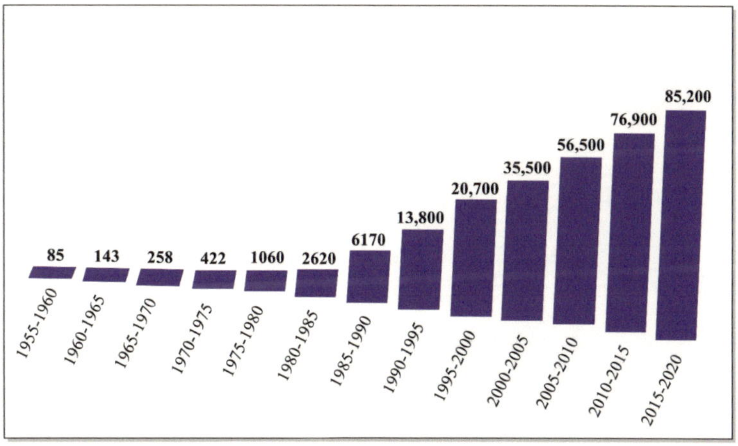

Figure 6.
Increasing number of publications on the CDs from the year 1955 to 2020.

substances of particular interest; (3) research explorations in the time period of 1935–1950; (4) the important period of maturation, i.e. 1950–1970, when notions become fully fledged, and in this period of time, the CDs had been structurally as well as chemically characterized, besides many more novel complexes were effectively studied; (5) historical landmarks—a period of diverse potential applications of CDs (i.e. since 1970s to the till date) in which the massive work has successfully been accomplished by a plethora of research groups globally. Importantly, at present, innumerable patents and research articles and books had already been published, and a lavish scientific literature has already been built up, accounting the wonderful chemistry of these beautifully simple yet much effectual supramolecular architectures. The research publications appeared in the scientific literature, in a time period of 1955–2020, are diagrammatically displaced in **Figure 6** [8].

3. Applications of CDs in drug delivery

Interestingly, cyclodextrins which are undoubtedly very effective complexing agents, and denoted with the different synonyms like cyclic oligosaccharides, cycloamyloses, cavitron, cycloglucan, cellulosine, and Schardinger sugars, are also invaluable from the drug delivery perspectives. These multipurpose CDs have also successfully found applications as drug delivery systems in nanoparticles, microcapsules, liposomes, oral drug delivery, nasal drug delivery, parenteral drug delivery, rectal drug delivery, peptide and protein delivery, dermal/transdermal delivery, and controlled drug delivery. Herewith, as can be inspected from **Table 2**, various approved and marketed drugs available in different countries worldwide are tabulated [9].

Name of drugs	Administration route	Trade name	Marketed in
α-Cyclodextrin			
Alprostadil (PGE1)	Intravenous	Prostavastin, Caverject, Edex	Europe, Japan, United States
Cefotiam hexetil HCl	Oral	Pansporin T	Japan
Limaprost	Oral	Opalmon, Prorenal	Japan
β-Cyclodextrin			
Benexate	Oral	Ulgut, Lonmiel	Japan
Dexamethasone	Derma	Glymesason	Japan
Iodine	Topical	Mena-Gargle	Japan
Nicotine	Sublingual	Nicorette	Europe
Nimesulide	Oral	Nimedex, Mesulid	Europe
Nitroglycerin	Sublingual	Nitropen	Japan
Omeprazole	Oral	Omebeta	Europe
Dinoprostone (PGE2)	Sublingual	Prostarmon E	Japan
Piroxicam	Oral	Brexin	Europe

Name of drugs	Administration route	Trade name	Marketed in
2-Hydroxypropyl-β-cyclodextrin			
Cisapride	Rectal	Propulsid	Europe
Hydrocortisone	Buccal	Dexocort	Europe
Indomethacin	Eye drops	Indocid	Europe
Itraconazole	Oral, intravenous	Sporanox	Europe, United States
Mitomycin	Intravenous	Mitozytrex	United States
Randomly methylated β-cyclodextrin			
17β-Oestradiol	Nasal spray	Aerodiol	Europe
Chloramphenicol	Eye drops	Clorocil	Europe
Sulphobutylether β-cyclodextrin			
Voriconazole	Intravenous	Vfend	Europe, United States
Ziprasidone maleate	Intramuscular	Geodon, Zeldox	Europe, United States
2-Hydroxypropyl-γ-cyclodextrin			
Diclofenac sodium	Eye drops	Voltaren	Europe

Table 2.
Approved and marketed drug-CD complexes in different markets worldwide.

4. Concluding remarks with future perspective

In this way, the chemistry of CDs is fully ripped, but there are always rooms to be occupied for newer advancements. I personally believe that the CDs will continue to garner the deepest interest from the scientific community across the world for several years in future, and that the newer potential applications of the CDs have yet to be exposed.

Remarkably, in the past few decades, the CDs have been catapulted into the distinction due to their enzyme mimic, catalysis, drug encapsulation, complexation, and molecular recognition behavior, etc. Moreover, they have also been attractively involved in the purification, polymerization, stabilization of the products, chemical treatment, food preservation, and other industrial processes. Finally, because of their chiral nature, selective modifications in their structures might further exploit their potential uses in modern asymmetric synthesis, molecular switches, molecular recognition, chiral separations, etc. Last but not least, because of their nontoxic character in addition to having the capabilities of complex formation with a varied vitamins, flavors, essential oils, perfumes, etc., they definitely have a gifted future in health-related products as well as biodegradable materials.

Author details

Rashid Ali
Organic and Supramolecular Functional Materials Research Laboratory, Department of Chemistry, Okhla, New Delhi, India

*Address all correspondence to: rali1@jmi.ac.in

References

[1] Crini G. A history of cyclodextrins. Chemical Reviews. 2014;**114**(21):10940-10975

[2] Khan AR, Forgo P, Stine KJ, Souza VT. Methods for selective modifications of cyclodextrins. Chemical Reviews. 1998;**98**:1977-1996

[3] Sharma N, Baldi A. Exploring versatile applications of cyclodextrins: An overview. Drug Delivery. 2016;**23**(3):729-747

[4] Gattuso G, Nepogodiev SA, Stoddart JF. Cyclodextrin-based catenanes and rotaxanes. Chemical Reviews. 1998;**98**:1959-1976

[5] Morin-Crini N, Fourmentin S, Fenyvesi E, Lichtfouse E, Torri G, Fourmentin M, et al. 130 years of cyclodextrin discovery for health, food, agriculture, and the industry: A review. Environmental Chemistry Letters. 2021;**19**:2581-2617

[6] Crini G. The contribution of Franz Schardinger to cyclodextrins: A tribute on the occasion of the centenary of his death. Journal of Inclusion Phenomena and Macrocyclic Chemistry. 2020;**97**:19-28

[7] Szejtli J. Introduction and general overview of cyclodextrin chemistry. Chemical Reviews. 1998;**98**:1743-1753

[8] Poulson BG, Alsulami QA, Sharfalddin A, Agammy EFE, Mouffouk F, Emwas A-H, et al. Cyclodextrins: Structural, chemical, and physical properties, and applications. Polysaccharides. 2022;**3**:1-31

[9] Davis ME, Brewster ME. Cyclodextrin-based pharmaceutics: Past, present and future. Nature Reviews. Drug Discovery. 2004;**3**(12):1023-1035

Chapter 2

Cyclodextrins: An Overview of Fundamentals, Types, and Applications

Rimsha Yousaf, Fizza Abdul Razzaq, Sajid Asghar, Muhammad Irfan, Ikram Ullah Khan and Syed Haroon Khalid

Abstract

Cyclodextrins are one of the most interesting pharmaceutical excipients with substantial theoretical and applied impacts in pharmaceutical industry. Even though the chemical foundation of these macrocyclic molecules was laid more than 100 years ago by Villiers and Schardinger, it was not until recently that cyclodextrins have been regarded as a subject of numerous potential pharmaceutical applications including inclusion complexation. This particular chapter discusses the fundamental concepts of cyclodextrin chemistry, structure, properties, and host-guest interaction with a special focus on molecular dynamics. Further in this regard, applications of cyclodextrins and numerous drug delivery approaches including novel lipid-based nanosystems are also highlighted.

Keywords: cyclodextrins, properties, solubility, drug molecules, inclusion complexes

1. Introduction

In general, cyclodextrins (CDs) comprise sugar molecules, which are combined together in the form of rings. The sugar molecules that specifically constitute cyclodextrins are "Glucopyranosides," that is, glucose molecules arranged in pyranose configuration. The first indication to a substance that is eventually identified as cyclodextrin was reported in 1891 after Villiers isolated a crystalline substance while working on enzymatic digestion of starch. After Villiers, Schardinger studied these crystalline substances and described the essentials of cyclodextrin molecules known to us in detail [1].

Chemically, cyclodextrin molecules are cyclic oligosaccharides consisting of alpha-1→4 linked D-glucopyranose units. Depending upon the number of glucopyranosides, cyclodextrin molecules can be categorized as alpha α (6), beta β (7), and gamma γ (8) cyclodextrins, respectively (**Figure 1**) [3].

As complicated as they sound, the CDs are comparatively easy to constitute. Like discussed above, cyclodextrins are typically obtained by treating starch with a variety of enzymes notably amylase or glucosyl transferases. Similar to the enzymes, sources of starch can also be variable resulting in particular ratio of α, β, and γ cyclodextrins [4].

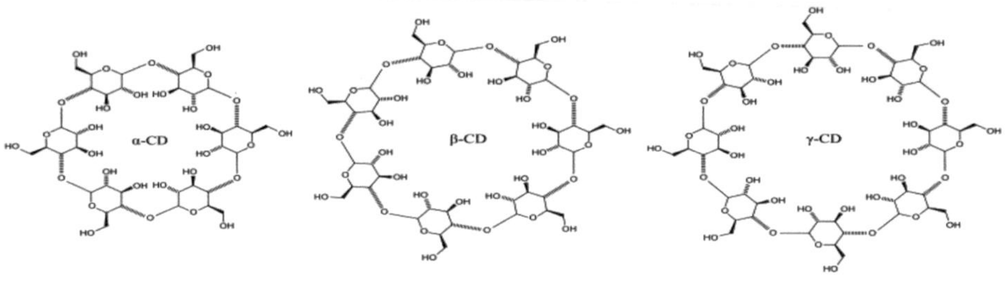

Figure 1.
Chemical structure of the α-, β-, and γ-cyclodextrins [2].

Furthermore, the CD molecules are large, but owing to stoichiometric constraints, it is not possible to acquire smaller CD molecules having less than six glucopyranosides residues. In contrast, those with higher glucose residues have been reported though concerns such as poor yield, and limited complexing ability renders them unacceptable for pharmaceutical use [5].

1.1 Structure and properties of cyclodextrin

In terms of structure, CDs have basket or truncated cone-like structure in which diameter of the inner cavity is a function of the glucopyranose units as shown in **Figure 2** [8]. The spherical arrangement of glucose units with secondary OH groups on wider end of the rim and primary OH groups on narrower end of the rim imparts it basket-like shape since the ability of primary OH groups to freely rotate decreases the diameter of the cavity at one end. Moreover, H-atoms bonded to CH group as well as OH groups form the external and hydrophilic exterior surface. In comparison to rims, internal cavity presents the hydrophobic microenvironment as it is surrounded with carbon and ether oxygen [9].

As cited earlier, there are three kinds of cyclodextrin, that is, Alpha α, Beta β, and Gamma γ also known as first or parent generation CDs. Due to the presence of sugar backbone in their framework, they can also be identified as cycloamyloses or dextrins [10, 11]. As far as the physicochemical properties are concerned, all the cyclodextrin molecules are large, hydrophilic, stable in basic media, hydrolyzable in acidic media,

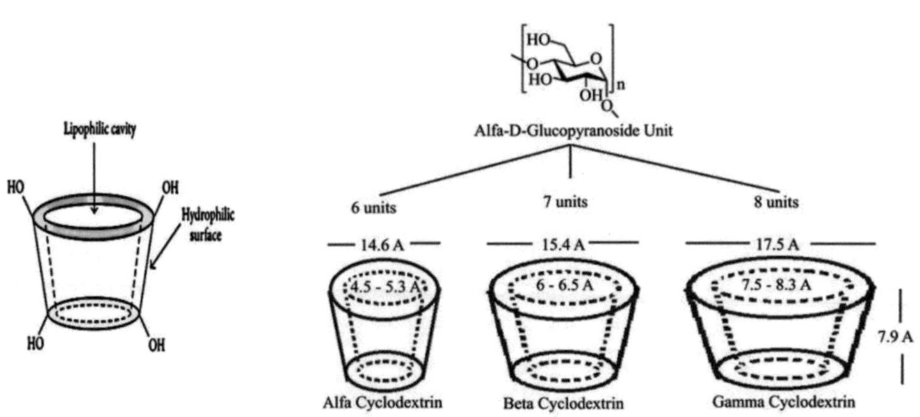

Figure 2.
Schematic diagram of shape and dimensions of parent cyclodextrin [6, 7].

Property	Alpha	Beta	Gamma
Glucose subunits Synonyms Height	Hexa Cyclo-hexaamylose (alfadex) 7.9	Hepta Cyclo-heptaamylose (betadex) 7.9	Octa Cyclo-octaamylose (gammadex) 7.9
Cavity diameter	4.5-5.3	6-6.5	7.5-8.3
External diameter	14.6	15.4	17.5
Solubility	14.5	1.85	23.2
Molecular weight	972	1135	1297

Table 1.
Important characteristics of parent cyclodextrin [7].

and similar in their ability to modify the physical, chemical, and biological features of drugs by yielding inclusion complexes (**Table 1**) [12].

The aqueous solubility of natural CDs and their complexes are known to be restricted, especially in case of β-CD, despite the fact that they are hydrophilic. This is owed to comparatively strong molecular bonding in CDs in their crystal state. Furthermore, significant number of intermolecular hydrogen bonds between secondary OH groups in the β-CD structure makes it stiff and inhibits the overall hydration. Interestingly, substituting any OH-group even by, for example, methyoxy group, can dramatically enhance the solubility. Nonetheless being less bulky, parent cyclodextrins exhibit lower molecular weight relative to their derivatives [13].

1.2 Derivatives of the CDs

Given the lower aqueous solubility, numerous scientists tried to prepare and evaluated a variety of derivatives of the CDs of medicinal interest. The CDs derivatives can be produced by polymerizing or substituting the methyl, carboxymethyl, ethyl, hydroxyethyl, sulfabutyl, or even saccharides. Bonding various functional groups causes chemical alterations into the main and secondary OH groups of the parent CD molecules [13, 14]. These derivatizations are carried out to achieve the following goals:

- To improve the solubility.
- To enhance the host-guest association or fitting.
- To stabilize the guest and lessen its reactivity and movement.

Remarkably, till date, a myriad of CD derivatives have successfully been produced and analyzed, yet only a small number including methylated, hydroxy alkylated, and ether substituted derivatives have been employed in studies involving novel pharmaceutical applications [15].

1. Methylated derivatives can be prepared by randomly methylating any secondary OH group in C2, C3, or C6 locations or by selectively methylating all secondary OH groups of C2 position and primary OH groups of C6. In relation to the natural CDs, methylated ones exhibit altered physical, chemical, and structural characteristics. Solubility of methylated CD is also substantially greater;

however, the solubility is inversely proportional to the temperature as it diminishes when temperature rises.

2. Similarly, hydroxyl alkyl derivatives also offer higher aqueous solubility and are one of the most extensively used derivative group. Preparation of hydroxy alkyl derivatives typically entails non-selective condensation of hydroxy alkyl agents in basic medium.

3. In contrast, ether derivatives reduce the solubility when OH groups are substituted by alkyl groups through ester or ether bond(s) [16, 17].

2. Effect of CDs on formulation properties

The most prominent attribute of the CDs is inclusion complexation, that is, the capability to allow a therapeutic agent or more characteristically just the hydrophobic portion of medicinal moiety into their internal cavity [18].

2.1 Mechanism of inclusion complexation

When water molecules are removed from the lipophillic cavity of cyclodextrins (which is in an energetically unfavorable environment due to the nature of the polar-polar interaction), the number of formed hydrogen bonds increases, and the repulsive interaction between guest and aqueous environment decreases, whereas the hydrophobic interaction increases as the guest molecule or lipophilic group with size, shape, and polarity compatible with the CD structure exerts itself in central cavity. As a result, a complex (**Figure 3**) is formed in an aqueous solution [19, 20].

When a complex is formed, covalent bonds are neither formed nor broken, and the drug molecules of complex as well as those of solution are in equilibrium. The capacity of guest to interact well with the host molecules to create a stable complex determines the binding strength of thus formed complex. Other factors involved in affecting this host-guest complexation mechanism are:

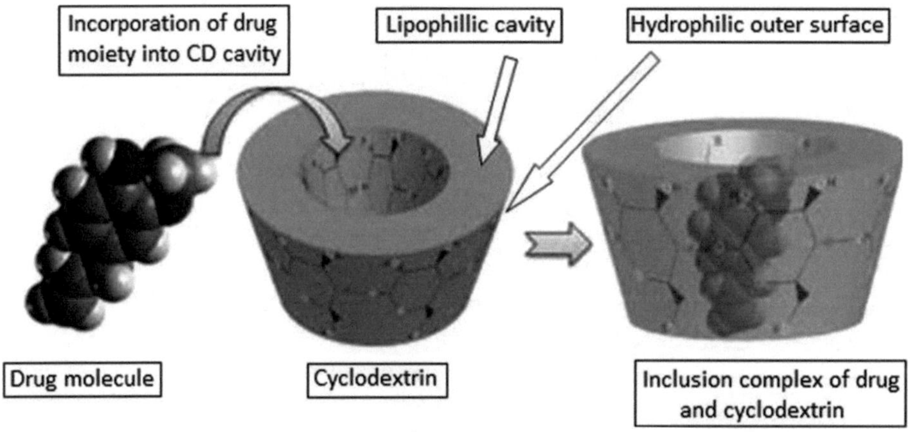

Figure 3.
Schematic illustration of the drug-CD inclusion complexation [2].

1. Steric factor—which is based on the proportion of the CD to drug size and/or on specific functional group of active component. If the active component is too big, it will not fit inside the cavity adequately. Moreover, the dimensions of these molecules also important in this aspect on the basis of dimension smaller compounds or those having aliphatic chain will form complex with the alpha cyclodextrin and those having higher molecular weight, for example, steroids will be accommodated by gamma cyclodextrin. On the other hand, heterocyclic as well as aromatic compounds will form complex with the β-CDs [21].

2. Besides the steric, thermodynamic interaction among various CD components, the host molecule is another important determinant. In order for complexation to take place, a favorable driving force that can draw the host molecule into the CD cavity is necessary. This thermodynamic force is attributable to the unique toroid or cone-like structure of the CDs [2].

3. Structure of substitution added to derivate CDs.

4. Places, where substitution are made within the molecule.

5. Finally, the degree of substitution

2.2 Types of complexes

It is crucial to measure the stability or dissociation constant of complexes, since they are an indicator of how a compound's physicochemical characteristics change after inclusion. The phase solubility method proposed by Connors and Higuchi [22] is most extensively used analytical procedure in this regard as shown in **Figure 4**, which is known as phase solubility profile.

Connors and Higuchi categorized complexes by examining the influence of solubilizer or ligand concentration on drug or substrate solubility. If by increasing the solubilizer or ligand concentration, there is a rise in substrate solubility, then the solubility profile is said to be A-type. Three additional profiles, that is, A_P, A_L, and A_N make up A-type profile. The A_L profile shows that solubility increases linearly with

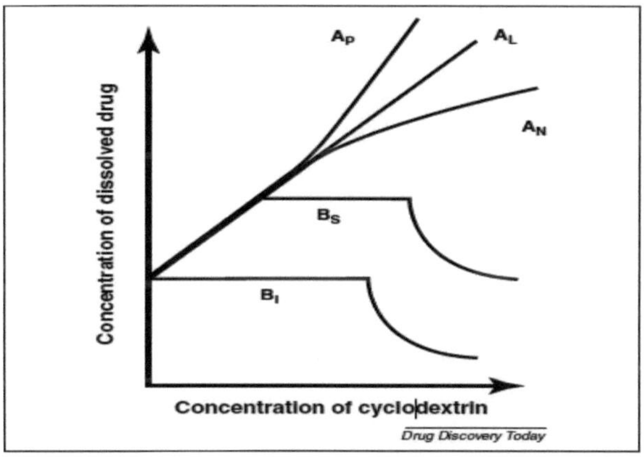

Figure 4.
A and B-type phase solubility profile and their subtypes [23].

ligand concentration or solubilizer concentration. A_P type shows an isotherm with significant deviation but in a positive way implying proportionality but only at higher quantities. On the other hand, A_N profile also exhibits deviation but is a negative way implying no proportionality even at higher amounts. All three trends together show that complexes formed are water-soluble. Type B profile on the other hand indicates that the complex has restricted solubility. They are typically seen with parent CDs especially with the β-CD. They are also of two types, that is, B_S and B_T [24].

Although inclusion complexes are expected to make up the majority of CD complexes but non-inclusion complexes (complex aggregate) that can dissolve drug through micelle formation can also form [2].

3. Applications of cyclodextrins

The exact potential of the CDs in the sector of pharmaceutical applications is because of their capability to affect several properties influencing the behavior and therapeutic outcomes of drugs. Cyclodextrins are typically employed to improve the solubility, permeability, stability as well as adverse effects including irritation. Generally, most of these applications are associated to their capability to form inclusion complexes.

3.1 Solubility and dissolution enhancement

The most extensive use of the CDs is to improve the solubility of drug in aqueous solutions. An increase in solubility also aids in improving bioavailability and hence therapeutic efficiency. Cyclodextrins have the capability to form the inclusion complexes, which increases the solubility and dissolution of drug molecules in the solid state [25]. Even though solubilization effects of all the CD molecules can be found throughout the literature, methylated CDs have the greatest potential of increasing the solubility as they decrease the crystallinity of drugs, which also increases the dissolution [26]. Although the influences of the CD complexation on it are extremely empirical, yet a number of historical findings permit several inferences:

- Firstly, the poorer the water solubility of the drug, the superior the solubility enhancement through the CD complexation.
- Secondly, derivatization of the CDs with reduced molar substitution offers the better solubilization than derivatives with more molar substitutions.
- Thirdly, solubilizing ability of the CDs is entirely dependent on the charge proximity to the cavity. Farther the proximity, better the ability.
- And lastly, it is feasible to increase complexation and consequently solubilization by incorporating various polymers of the group.

3.2 Permeability across biological membrane

Remarkably, permeability across the biological membranes can be affected by several factors, which include molecular weight, molecular structure in addition to the partition coefficient. CDs have no part in increasing the permeation of hydrophilic drugs, but in CD complexation, free drug has the affinity to penetrate biological membrane [27]. At this point, delivery across biological membrane is entirely dependent on the drug formulation as well as the barrier. Delivery across barrier controlled by water diffusion layers can be affected by cyclodextrins, but those across lipophilic membranes are limited. The only exception to this barrier is inclusion of hydrophobic cyclodextrins that can effortlessly cross the mucosa [28].

3.3 Higher photo- and thermal stability

Another important signature of these excipients is their influence on the chemical stability of pharmaceuticals. Whenever any drug formulation has to be developed, stability parameters and the factors affecting the stability parameters should be kept in mind, and appropriate stability enhancers should be added as per the requirements. CDs are widely known for their capability to reduce the effect of temperature, light, and oxygen, thereby increasing the overall stability [29, 30]. Degradation of the product in the presence of light can lead to the several adverse effects. Higher photo stability was found when a complex of CD and vitamin E was formed. Apart from the protective effect of CD stability, studies are also important to discover the degree to which any formulation can be prevented from the excipient mediated degradation [31].

3.4 Improved drug safety

When CDs increase the solubility, dissolution, and bioavailability of the drugs [28], it means that drug will have the required residence time in the body and will not stay longer, thus reducing the risks of toxic effects [32]. A research was conducted on an anti-viral drug ganciclovir combined with CD, and it was found that toxic effects of the drug were reduced and efficacy was significantly improved. Similarly, irritation caused by both intravenous and ophthalmic products can also be reduced by CDs [33].

3.5 Control of drug release

CDs having ethyl group and acyl group have the potential to prolong drug release [34]. One alternative for controlling drug release is to utilize the epithelial surface of GIT in which per-Obutanoyl β-CD is known for its mucoadhesive property. HP-β-CDs are being utilized for their gel forming property, thus can extend the release of drug. In controlled drug delivery systems, osmotic pumps are widely utilized as they are unique and provide the uniform concentration of drug in the systemic circulation [35]. Advanced forms of extended delivery systems can be developed by joining the CD conjugates with respective release formulations. This effect was seen when ketoprofen having β-CD was combined with ketoprofen, and this formulation was added in CD conjugates, which provided a repeated release profile [36, 37].

4. Drug delivery approaches

4.1 Drug delivery by oral route

Drug delivery by oral route has traditionally been the most prevalent option for designing delivery systems. Drug release in oral delivery system may be controlled by dissolution, diffusion, pH, or osmosis [38]. The usage of CDs in an oral delivery system is to increase the rate at which dissolution occurs—forming inclusion complexes with CD aids in increasing the solubility of drugs and hence transport of drugs across aqueous phase to lipid membrane in GIT [39]. The hydrophobic derivatives of CDs are mostly employed to accomplish this goal. In case of buccal and sublingual routes, rapid increase in drug concentration can also be achieved by complexation; however, in order to exhibit the therapeutic effect, drug must need to get released from the complex. For sublingual route, it is a little bit difficult since the amount of saliva as well as contact time is limited [40].

Cyclodextrins especially hydrophobic ones, that is, ethylated CDs, are also very important in achieving site-specific or sustained drug release. Additionally, cyclodextrins have productively been utilized in matrix tablets as well as osmotic pumps to control the drug release [41].

4.2 Ocular drug delivery

The primary treatment of an ocular aliment is mainly topical application of drug as aqueous solution. The current findings ascertain that cyclodextrin molecules are helpful components in ocular preparations, since they can enhance the solubility, stability, and consequently bioavailability of the ophthalmic formulations [42]. Among the CDs, hydrophilic cyclodextrins, mainly SB β-CD and HP β-CD, are reported to be most compatible and nontoxic [43]. It is well known that only a small amount an ophthalmic drug can actually reach systemic circulation, but increasing the availability of a drug at corneal surface through the CD complexation can easily enhance ocular bioavailability of hydrophobic drugs [44].

4.3 Nasal drug delivery

In order to have systemic absorption, drug must have optimum solubility in nasal fluids. Moreover, an optimum nasal formulation also must not have any effect on the defensive functions of cilia in respiratory tract. Both hydrophilic and hydrophobic CDs are the highly employed in this regard, as they can enhance the solubilization as well as the permeation, correspondingly. Besides, they are highly effective in small concentration and stereotypically inert from toxicological perspective [45, 46].

4.4 Transdermal drug delivery

Stratum cornea serves as the main barrier in the delivery of drugs through the skin. Various penetration enhancers are often employed to enhance the delivery across the barrier. Owing to the hydrophobic properties, cyclodextrins have the ability to deliver across water diffusion layer; however, if absorption is dependent on the lipophilic barrier solely, CDs are unable to deliver the drug dermally. Therefore, suitable selection of an aqueous vehicle is highly important [32, 47].

4.5 Novel drug delivery

Captivatingly, CD and its derivatives have been employed to develop the novel systems having supramolecular architectures such as micelles [48], nanosponges [49], nanoparticles [50], nanovesicles [51], etc., to build the functional platforms. Among these delivery systems, lipid nanocarriers are arguably the most common nanomaterials, which are used in association with modified CDs. Being biodegradable and biocompatible, these systems offer versatile advantages including targeted delivery, stability, and co-drug loading (i.e., both hydrophobic and hydrophilic). In addition, they also exhibit superior efficacy and pharmacokinetics [52]. Conducive studies on the lipid nanosystems including parent and derivated CDs have proven the suitability of this approach to enhance the bioavailability of numerous pharmaceutical formulations; hence, continually increasing its implications in different disorder, for instance, diabetes [53], hypertension [54], cancer [55], and many other ailments.

5. Conclusions

It is evident from the data given here that cyclodextrins due to their distinctive cone-like structure and unique properties can provide an excellent option to address various issues regarding drug and its delivery. CDs have been introduced successfully in pharmaceutical industry because of their unique property of forming dynamic inclusion complexes, thus improving the solubility, dissolution, bioavailability, and release profile of numerous drugs. The effectiveness of different CDs in improving the therapeutic potential of drugs depends upon both the type of guest molecule and the CD as they can influence types of the complexes formed. According to Connor and Higuchi, complexes can be A-type, B-type, and even non-inclusion type, such as complex aggregation.

Apart from traditional drug delivery system, recent research on CD-based nanosystems has grown as the CDs are biocompatible and can nicely serve as the platform for the formulation and pharmacokinetic efficiency without raising the risks.

Conflict of interest

The authors declare no conflict of interest.

Author details

Rimsha Yousaf[1], Fizza Abdul Razzaq[2], Sajid Asghar[2], Muhammad Irfan[2], Ikram Ullah Khan[2] and Syed Haroon Khalid[1*]

1 Faculty of Pharmaceutical Sciences, Government College University, Faisalabad, Pakistan

2 Department of Pharmaceutics, Faculty of Pharmaceutical Sciences, Government College University, Faisalabad, Pakistan

*Address all correspondence to: haroonkhalid80@gmail.com; syedharoonkhalid@gcuf.edu.pk

References

[1] Crini G. A history of cyclodextrins. Chemical Reviews. 2014;**114**(21): 10940-10975

[2] de Jesus MB et al. Non-inclusion complexes between riboflavin and cyclodextrins. Journal of Pharmacy and Pharmacology. 2012;**64**(6):832-842

[3] Singh M, Sharma R, Banerjee U. Biotechnological applications of cyclodextrins. Biotechnology Advances. 2002;**20**(5-6):341-359

[4] Alves-Prado H et al. Production of cyclodextrins by CGTase from Bacillus clausii using different starches as substrates. In: Biotechnology for Fuels and Chemicals. ABAB Symposium. Switzerland: Humana Press, Springer; 2007. pp. 123-133

[5] Biwer A, Antranikian G, Heinzle E. Enzymatic production of cyclodextrins. Applied Microbiology and Biotechnology. 2002;**59**(6):609-617

[6] Alanazi F, Salah D. Gold nanoparticles incorporated with Cyclodextrins and its applications. Journal of Biomaterials and Nanobiotechnology. 2021;**12**(4):79-97

[7] Banchero M, Ronchetti S, Manna L. Characterization of ketoprofen/methyl-β-cyclodextrin complexes prepared using supercritical carbon dioxide. Journal of Chemistry. 2013;**2013**:1-8

[8] Dodziuk H, Nowiński K. Structure of cyclodextrins and their complexes: Part 2. Do cyclodextrins have a rigid truncated-cone structure? Journal of Molecular Structure: THEOCHEM. 1994;**304**(1):61-68

[9] Irie T, Uekama K. Pharmaceutical applications of cyclodextrins. III. Toxicological issues and safety evaluation. Journal of Pharmaceutical Sciences. 1997;**86**(2):147-162

[10] Inoue Y. NMR studies of the structure and properties of cyclodextrins and their inclusion complexes. In: Annual Reports on NMR Spectroscopy. Amsterdam: Elsevier; 1993. pp. 59-101

[11] Baglole KN, Boland PG, Wagner BD. Fluorescence enhancement of curcumin upon inclusion into parent and modified cyclodextrins. Journal of Photochemistry and Photobiology A: Chemistry. 2005;**173**(3):230-237

[12] Kelly RM, Dijkhuizen L, Leemhuis H. The evolution of cyclodextrin glucanotransferase product specificity. Applied Microbiology and Biotechnology. 2009;**84**(1):119-133

[13] Loftsson T, Brewster ME. Pharmaceutical applications of cyclodextrins: Basic science and product development. Journal of Pharmacy and Pharmacology. 2010;**62**(11):1607-1621

[14] Szente L, Szejtli J. Highly soluble cyclodextrin derivatives: Chemistry, properties, and trends in development. Advanced Drug Delivery Reviews. 1999;**36**(1):17-28

[15] Stella VJ, He Q. Cyclodextrins. Toxicologic Pathology. 2008;**36**(1): 30-42

[16] Croft AP, Bartsch RA. Synthesis of chemically modified cyclodextrins. Tetrahedron. 1983;**39**(9):1417-1474

[17] Ogawa N et al. Quetiapine free base complexed with cyclodextrins to improve solubility for parenteral use. Chemical and Pharmaceutical Bulletin. 2013;**61**(8):809-815

[18] Liu L, Guo Q-X. The driving forces in the inclusion complexation of cyclodextrins. Journal of Inclusion Phenomena and Macrocyclic Chemistry. 2002;**42**(1):1-14

[19] MosherG,ThompsonD.Complexation and cyclodextrins. Encyclopedia of Pharmaceutical Technology. 2002;**19**:49-88

[20] Nuchuchua O et al. Physicochemical investigation and molecular modeling of cyclodextrin complexation mechanism with eugenol. Food Research International. 2009;**42**(8):1178-1185

[21] Jacob S, Nair AB. Cyclodextrin complexes: Perspective from drug delivery and formulation. Drug Development Research. 2018;**79**(5):201-217

[22] Loftsson T, Másson M, Brewster ME. Self-association of cyclodextrins and cyclodextrin complexes. Journal of Pharmaceutical Sciences. 2004;**93**(5): 1091-1099

[23] BrewsterME,LoftssonT.Cyclodextrins as pharmaceutical solubilizers. Advanced Drug Delivery Reviews. 2007;**59**(7):645-666

[24] Pápay ZE et al. Comparative evaluation of the effect of cyclodextrins and pH on aqueous solubility of apigenin. Journal of Pharmaceutical and Biomedical Analysis. 2016;**117**:210-216

[25] Becket G, Schep LJ, Tan MY. Improvement of the in vitro dissolution of praziquantel by complexation with α-, β-and γ-cyclodextrins. International Journal of Pharmaceutics. 1999;**179**(1):65-71

[26] Londhe V, Nagarsenker M. Comparison between Hydroxypropyl b-Cyclodextrin and Polyvinylpyrrolidone As carriers for carbamazepine solid Despersions. Indian Journal of Pharmaceutical Sciences. 1999;**61**(4):237

[27] Laza-Knoerr A, Gref R, Couvreur P. Cyclodextrins for drug delivery. Journal of Drug Targeting. 2010;**18**(9):645-656

[28] Challa R et al. Cyclodextrins in drug delivery: An updated review. AAPS PharmSciTech. 2005;**6**(2):E329-E357

[29] Dhall M, Madan A. Comparison of cyclodextrins and urea as hosts for inclusion of drugs. Journal of Inclusion Phenomena and Macrocyclic Chemistry. 2017;**89**(3):207-227

[30] Budhwar V. Cyclodextrin complexes: An approach to improve the physicochemical properties of drugs and applications of cyclodextrin complexes. Asian Journal of Pharmaceutics (AJP): Free full text articles from Asian J Pharm. 2018;**12**(02):S394-409

[31] Popielec A, Loftsson T. Effects of cyclodextrins on the chemical stability of drugs. International Journal of Pharmaceutics. 2017;**531**(2):532-542

[32] Rasheed A. Cyclodextrins as drug carrier molecule: A review. Scientia Pharmaceutica. 2008;**76**(4):567-598

[33] Ukema K, Hirayama F, Irie T. Cyclodextrin drug carrier system. Chemical Reviews. 1998;**98**:2045-2076

[34] Memişoğlu E et al. Amphiphilic β-cyclodextrins modified on the primary face: Synthesis, characterization, and evaluation of their potential as novel excipients in the preparation of nanocapsules. Journal of Pharmaceutical Sciences. 2002;**91**(5):1214-1224

[35] Vyas A, Saraf S, Saraf S. Cyclodextrin based novel drug delivery systems. Journal of Inclusion

Phenomena and Macrocyclic Chemistry. 2008;**62**(1):23-42

[36] Uekama K. Design and evaluation of cyclodextrin-based drug formulation. Chemical and Pharmaceutical Bulletin. 2004;**52**(8):900-915

[37] Kamada M et al. Cyclodextrin conjugate-based controlled release system: Repeated-and prolonged-releases of ketoprofen after oral administration in rats. Journal of Controlled Release. 2002;**82**(2-3):407-416

[38] Fine-Shamir N et al. Toward successful cyclodextrin based solubility-enabling formulations for oral delivery of lipophilic drugs: Solubility–permeability trade-off, biorelevant dissolution, and the unstirred water layer. Molecular Pharmaceutics. 2017;**14**(6):2138-2146

[39] Loftsson T, Brewster ME, Masson M. Role of cyclodextrins in improving oral drug delivery. American Journal of Drug Delivery. 2004;**2**(4):261-275

[40] Tiwari G, Tiwari R, Rai AK. Cyclodextrins in delivery systems: Applications. Journal of Pharmacy and Bioallied Sciences. 2010;**2**(2):72

[41] Okimoto K et al. Design and evaluation of an osmotic pump tablet (OPT) for prednisolone, a poorly water soluble drug, using (SBE) 7m-β-CD. Pharmaceutical Research. 1998;**15**(10):1562-1568

[42] Kim E-Y et al. rhEGF/HP-β-CD complex in poloxamer gel for ophthalmic delivery. International Journal of Pharmaceutics. 2002;**233**(1-2):159-167

[43] Pahuja P, Kashyap H, Pawar P. Design and evaluation of HP-β-CD based voriconazole formulations for ocular drug delivery. Current Drug Delivery. 2014;**11**(2):223-232

[44] Loftssona T, Järvinen T. Cyclodextrins in ophthalmic drug delivery. Advanced Drug Delivery Reviews. 1999;**36**(1):59-79

[45] Loftsson T et al. Cyclodextrins in drug delivery. Expert Opinion on Drug Delivery. 2005;**2**(2):335-351

[46] Mianowska B, Śliwa W. Cyclodextrins as drug carriers. Prace Naukowe. Chemia i Ochrona Środowiska/Akademia im. Jana Długosza w Częstochowie. 2008;**12**:155-162

[47] Chen Y, Wang M, Fang L. Biomaterials as novel penetration enhancers for transdermal and dermal drug delivery systems. Drug Delivery. 2013;**20**(5):199-209

[48] Liu Y et al. Bioadhesive chitosan-coated cyclodextrin-based superamolecular nanomicelles to enhance the oral bioavailability of doxorubicin. Journal of Nanoparticle Research. 2014;**16**(10):1-15

[49] Cavalli R, Trotta F, Tumiatti W. Cyclodextrin-based nanosponges for drug delivery. Journal of Inclusion Phenomena and Macrocyclic Chemistry. 2006;**56**(1):209-213

[50] Krauland AH, Alonso MJ. Chitosan/cyclodextrin nanoparticles as macromolecular drug delivery system. International Journal of Pharmaceutics. 2007;**340**(1-2):134-142

[51] Dhule SS et al. Curcumin-loaded γ-cyclodextrin liposomal nanoparticles as delivery vehicles for osteosarcoma. Nanomedicine: Nanotechnology, Biology and Medicine. 2012;**8**(4):440-451

[52] Gharib R et al. Liposomes incorporating cyclodextrin–drug inclusion complexes: Current state of knowledge. Carbohydrate Polymers. 2015;**129**:175-186

[53] Razzaq FA et al. Glimepiride-loaded Nanoemulgel; development, In vitro characterization, ex vivo permeation and In vivo antidiabetic evaluation. Cell. 2021;**10**(9):2404

[54] Mariangela de Burgos MD et al. New formulation of an old drug in hypertension treatment: The sustained release of captopril from cyclodextrin nanoparticles. International Journal of Nanomedicine. 2011;**6**:1005

[55] Sett R et al. Effect of temperature and salts on niosome-bound anti-cancer drug along with disruptive influence of cyclodextrins. Spectrochimica Acta Part A: Molecular and Biomolecular Spectroscopy. 2020;**234**:118261

Chapter 3

Cyclodextrin-Based Sensors for the Recognition of Small Molecules

Ishfaq Ahmad Rather, Ahmad Hasan and Rashid Ali

Abstract

Owing to the selective recognition ability, exceptional biocompatibility, water solubility, non-toxicity, economically inexpensive, commercial availability, and easy functionalization, cyclodextrins (CDs) act as the main building blocks for the creation of beautifully simple yet much effective supramolecular architectures of fundamental interest. Over the past few decades, CDs have engrossed a noteworthy interest in the scientific community because of their usage in the development of chemical sensors *via* molecular recognition phenomenon. Bearing the delightful sensing capability of CDs in mind, herewith, we envisioned to disclose the recent developments in the sensing of diverse biologically significant small molecules by CDs through colorimetric, fluorescence, electrochemical, and potentiometric response. Sensing events and corresponding distinguishing optical features in cyclodextrin-based monomers, dimers, clusters, and nano-assemblies have been elaborated in detail. The authors are of the opinion that this chapter will offer new dimensions to supramolecular sensors in general and CDs-based sensors in particular.

Keywords: cyclodextrins, host-guest interaction, molecular recognition, small molecules, colorimetric and electrochemical sensors

1. Introduction

Basically, the host-guest non-covalent interaction is the major subtopic of supramolecular chemistry, which succors us to realize the recognition of guest entities, particularly through non-covalent supramolecular interactions [1, 2]. In recent years, the supramolecular host molecules, such as cyclophanes, crown ethers, cryptands, calix[n]arenes, calix[n]pyrroles, cucurbiturils, and cyclodextrins, have drawn an enormous interest of the scientific community worldwide because of their exceptional signatures, particularly molecular recognition and sense of specific analytes, and still much new chemistry with these old macrocycles is to be explored [3–7]. Among the above-mentioned host architectures, the naturally occurring cyclodextrins (CDs) are regarded as most essential by virtue of their selective recognition capability, exceptional biocompatibility, water solubility, non-toxicity, economically inexpensiveness, commercial availability, and easy-functionalization [8]. With the aid of host-guest chemistry, the CDs have found a range of applications in various fields of science and technology *viz.* supramolecular self-assemblies, material sciences, pharmaceutical chemistry, biochemistry, polymer chemistry, electronics, catalysis, and nanotechnology, besides biotechnological and chemical industries [9–11]. Remarkably, CDs have also been employed as the bricks in building frequent supramolecular structures of particular interest, such as polyrotaxanes, rotaxanes, catenanes, and supramolecular polymeric materials [12].

The CDs are cyclic oligosaccharides-based seminatural products, mostly comprising of 6–8 units (α-, β-, & γ-cyclodextrins) of glucose connected through α-1,4-gycosidic linkages to generate the torus-shaped molecules portrayed by a hydrophilic surface and hydrophobic central cavity (**Figure 1**) [8]. Notably, CDs having glucose units less than six are too much strained for existence, while the CDs containing more than eight glucose units are readily soluble and very difficult to isolate. With an increase in the number of glucopyranose units from six to eight, the inner cavity diameter also increases from 0.44 to 0.83 nm. In particular, the inner cavity diameter of 0.44 nm in α-CD is suitable to capture benzene molecule, whereby β-CD (0.62 nm) holds an appropriate cavity to encapsulate the naphthalene molecule, and importantly, the γ-CD (0.83 nm), can easily occupy the larger guest molecules *viz.* fullerene [13, 14]. The shape of these CDs resembles like a bucket and hence offers a narrow and large entrance on opposite sides. Typically, it has been revealed that there exist primary OH-moieties on the side of narrow cavities and they have got recognition as a primary face. On the front, secondary OH moieties are present on the side of a large cavity and are generally dubbed as the secondary face. It is by virtue of these primary and secondary OH groups that these CDs are selective toward the inclusion of guest entities of particular importance. As a matter of the fact, both primary and secondary OH groups arrange themselves on the outer side of two recognized faces, and the whole inside cavity of these CDs becomes a hydrophobic microenvironment. The hydrophobicity of the inner cavity in turn is responsible for the inclusion of typically hydrophobic guests in aqueous media [8]. Importantly, over the passage of time, selective methods for ease functionalization of CD-scaffold are being proposed

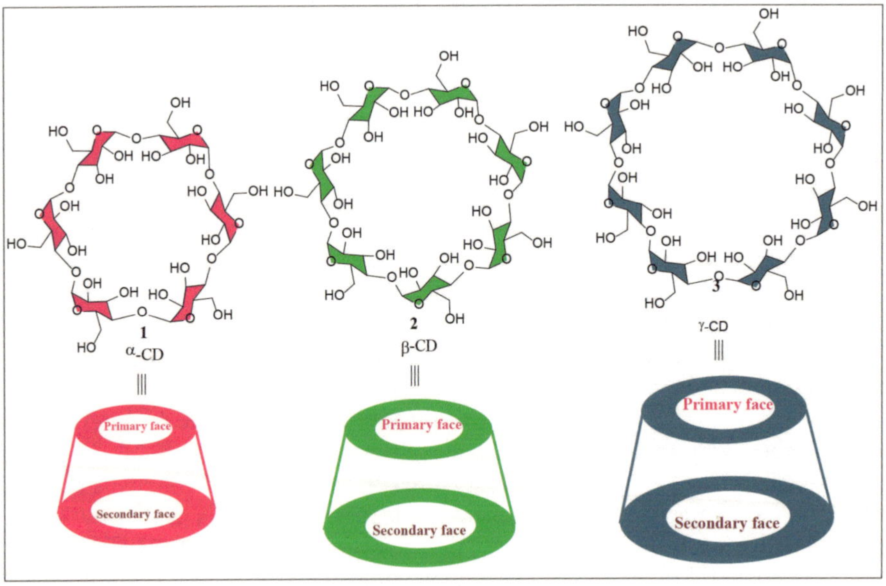

Figure 1.
Chemical structures and the 3D-pictorial representation of different CDs.

constantly in order to enhance the recognition properties of CD-based supramolecular hosts toward analytes [15].

As can be straightforwardly inspected from the scientific publications appearing in the literature, the domain of chemical sensors in general and CD-based sensors in particular are rapidly progressing and strengthening their roots in various aspects of our day-to-day life besides bringing a revolution in diverse arena of science and technology [16]. Keeping these facts in mind and also to expose the importance of sensory materials based on CDs; in this meticulous review chapter, we indented to highlight the recent developments in addition to the conceptual background of CD-based chemical sensors. Hopefully, the readers will enjoy this draft and will for sure be further to explore these old yet new types of macromolecular platforms to an advanced level.

2. Chromophore-appended cyclodextrins as classical chemical sensors

Among a variety of chemical sensors, the optical chemosensors are truly interesting and advantageous [17]. This can be ascribed to the fact that optical variations *viz.* color, absorption, and/or emission developing after the recognition of the targeted guest analyte by a typical host molecule are most of the time directly visible through naked eye. In these optical chemosensors, chromophores are attached with basic sensing scaffolds in order to utilize their optical variation, impending for the determination of successful recognition/sensing event [18].

Keeping in consideration, the fact that microenvironment of the utilized chromophores offer changes in color as well as fluorescence pattern, and to get optimum output, researchers globally have functionalized the CDs with a variety of dyes besides the fluorescent moieties [19, 20]. Using chromophore-appended CDs, the detection of a range of hydrophobic guest molecules inside the hydrophobic cavities has extensively been studied in recent years. However, a plethora of chromophore-appended CDs have been constructed and their sensing activities have also been accomplished by several research groups worldwide. But, the pioneering work in this field has been revealed by Ueno and teammates; they reported many CD-based fluorescent chemical sensors through the installation of diverse fluorophores (dansyl, pyrene, anthracene, *etc.*) in CDs *via* flexible linker [21, 22]. By means of this rigid spacer, no self-inclusion complex formation has been noticed. This in turn exposes the fluorophore to a hydrophilic environment and leads to the fluorescence quenching of the CDs. Consequently, the addition of hydrophobic guest leads to its inclusion in the hydrophobic cavity of CDs, thereby bringing the appended fluorophore to a more hydrophobic environment in comparison with the free state of CD-fluorophore conjugates. Hence, enhancement in the fluorescence leads to the "turn-on" fluorescence response (**Figure 2b**) [21].

On the other hand, the same group has also constructed various colorimetric indicator dyes (*viz.* phenolphthalein, methyl red, *p*-nitrophenol, alizarin yellow) appended CD-based chemosensors, wherein the dye moiety is included in the hydrophobic cavity of CDs and generates the self-inclusion complex well isolated from the exterior aqueous hydrophilic media [23–25]. In this self-inclusion complex state, the color changes of the appended dye moiety through protonation/deprotonation-assisted pH variation are suppressed (**Figure 3**). The consequent addition of competitive hydrophobic guest molecule leads to the segregation of appended dye moiety from the interior of CD hydrophobic cavity to the exterior hydrophilic environment.

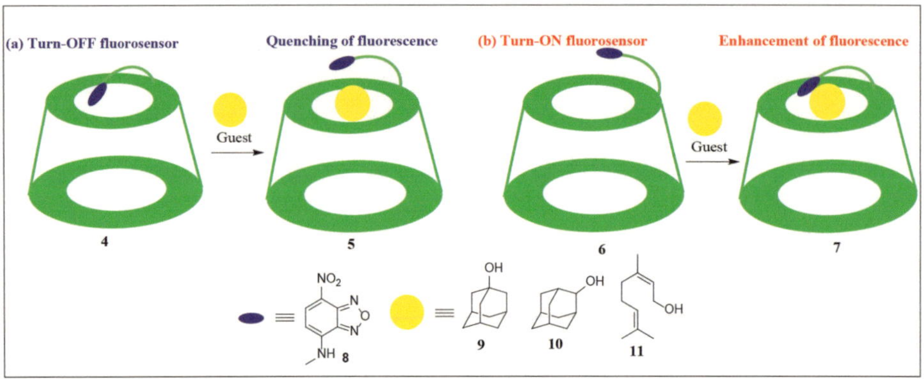

Figure 2.
Pictorial representation of the turn-off (a) and turn-on (b) fluorescent CD-based chemical sensors developed by Ueno and co-workers.

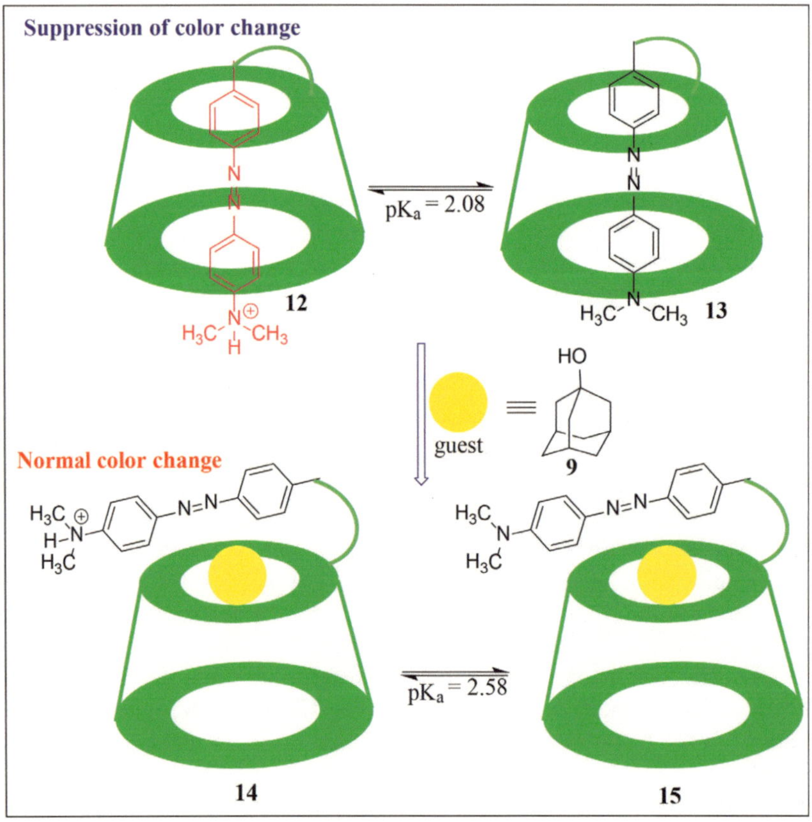

Figure 3.
Schematic representation of the p*-methyl red appended CD chemical sensor.*

In this manner, the appended dye moiety displays normal color variations upon changing the pH through the protonation or deprotonation tactic (**Figure 3**) [26, 27].

Sulfur dioxide is widely used as a preservative and antioxidant in the food and beverage industries. Thus, constructing sulfur dioxide sensors is of utmost significance in food and analytical chemistry. In this regard, Levine and co-workers have

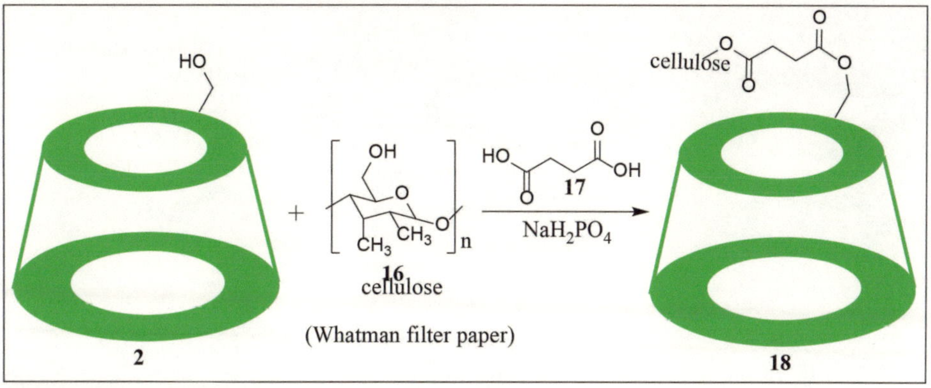

Figure 4.
Schematic depiction of chemical reaction involved in the attachment of β-CD (2) with Whatman filter paper.

modified the Whatman filter paper with *β*-CD (**2**) and manganese in order to develop a colorimetric sensor for sulfur dioxide in an aqueous solution (**Figure 4**) [28]. It has been revealed that the developed sensor is sensitive (limit of detection up to 33 ppm), practical, and broadly applicable in the rapid detection of sulfur dioxide *via* naked eye color change. Besides, the redox reaction of the manganese has been found responsible for the perceived naked eye color variations and other UV-Vis spectral variations. For practical applications, these studies pave the way toward the construction of CD-based novel sulfur dioxide sensors for their employment in beverage and food industries.

3. Metallocyclodextrin-based chemical sensors

The metallocyclodextrin-based CDs have been developed by various research groups and utilized in the field of chemical sensors [29, 30]. To this line, ligands consisting of the metal binding sites, for example, crown ether, diethylenetriamine-pentaacetate (DTPA), and ethylenediaminetetraacetate (EDTA), have successfully been reported. Noticeably, among the various metal ions, lanthanide metal ions (Eu^{2+} & Tb^{2+}) are primarily used in the fabrication of metallocyclodextrin-based chemical sensors by virtue of the fact that they exhibit strong fluorescence and also showed the longer lifetimes [31]. Out of various sensing mechanisms, absorption energy transfer emission (AETE) has been found responsible for the sensing of metallocyclodextrins. This sensing mechanism preliminary involves the excitation of light harvesting guest molecule *via* absorption of photon energy followed by the transfer of energy to a photoactive metal ion (Eu^{2+}/Tb^{2+}) and subsequent emission from these metal ions. It has been revealed that the complexes of Eu^{2+} and Tb^{2+} ions with the appended CDs *viz.* crown ether-CD (**19**) and DTPA-CD (**20**) conjugate display slight fluorescence in aqueous solution due to dearth of aromatic hydrocarbons acting as light-harvesting groups (**Figure 5**) [32]. However, the addition of aromatic guest/light-harvesting molecules, such as benzene, toluene, and biphenyl, leads to the inclusion of these molecules into the inner hydrophobic cavity of metallocyclodextrins, thereby displaying fluorescence enhancement *via* AETE, and, hence, offers a unique approach to develop the turn-on fluorescent chemical sensors (**Figure 5**). On the other hand, Reinhoudt and teammates have constructed *β*-CD dimer (**22**) in which two *β*-CD units

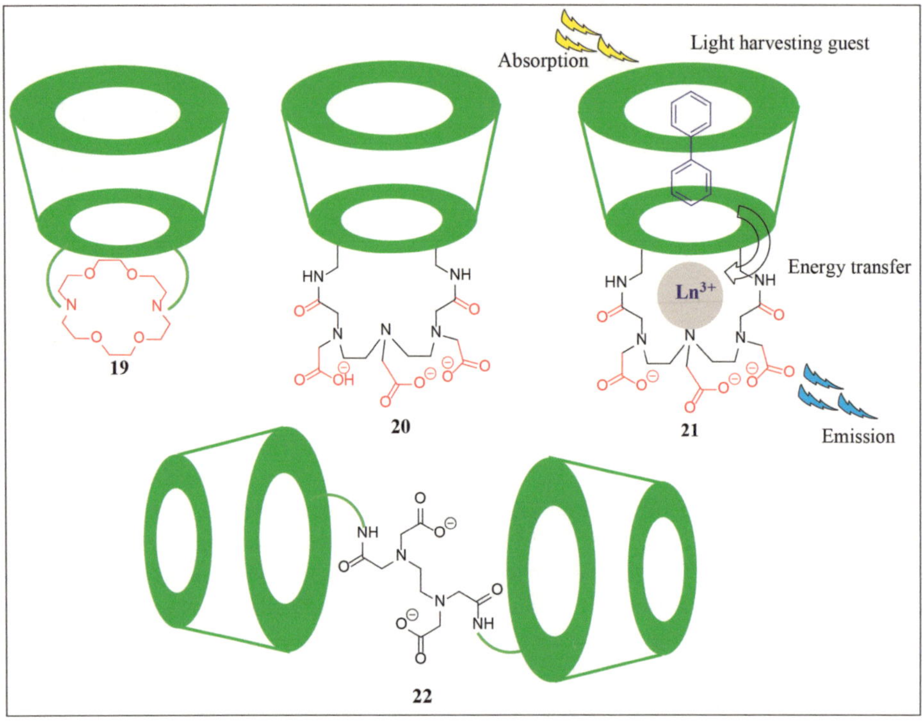

Figure 5.
*Diagrammatic illustration of crown ether appended CD (**19**), DTPA appended CD (**20**), and EDTA linked β-CD dimer (**22**). Moreover, the mechanistic overview of AETE sensing phenomenon involving light harvesting guest molecules is also shown.*

are linked through EDTA (**Figure 5**) [33]. From the experimental studies, it has been noticed that the complex formation occurs between **22** and lanthanide metal ions, which upon the addition of biphenyl-linked adamantane dimer results in the inclusion of adamantyl ends in the hydrophobic cavities of *β*-CD dimer (**22**). Consequently, AETE has been noticed from the biphenyl group of adamantane dimer to the lanthanide metal ion complexed with *β*-CD dimer (**22**). In this way, the overall lanthanide metallocyclodextrin-based assembly functions like a turn-on fluorescent chemical sensor. Additionally, polypyridine, as well as hepta bipyridine, appended CDs forms the complexes with lanthanide metal ions and acts as the chemical sensors toward targeted guest molecules *via* AETE sensing phenomenon [34, 35].

Liu *et al.* have studied the transition metal cation ligand-appended CDs as fluorescent chemical sensors [36]. They have synthesized *β*-CD dimer (**23**) in which two *β*-CD units are joined through the biquinolino subunits (**Figure 6**), and by virtue of this group, *β*-CD dimer (**23**) forms a complex with Cu(II) transition metal ion. Upon the resulting addition of steroid guest molecule, a 1:1 sandwich-type inclusion complex was formed, displaying the enhancement in fluorescence and hence acting as an efficient fluorescent chemical sensor. Moreover, the same group has also synthesized quinoline functionalized *β*-CD-based selective fluorescent sensor (**24**) for Zn(II) ion, among several other interfering metal ions, such as Ca(II) and Mg(II) (**Figure 6**). For real-world uses, the sensor (**24**) might prove highly valuable as an imaging agent for Zn(II) in living cells or tissues [37]. In a separate report, Yang *et al.* have established a selective and sensitive *β*-CD-based fluorescent sensor (**25**) for the recognition of Zn

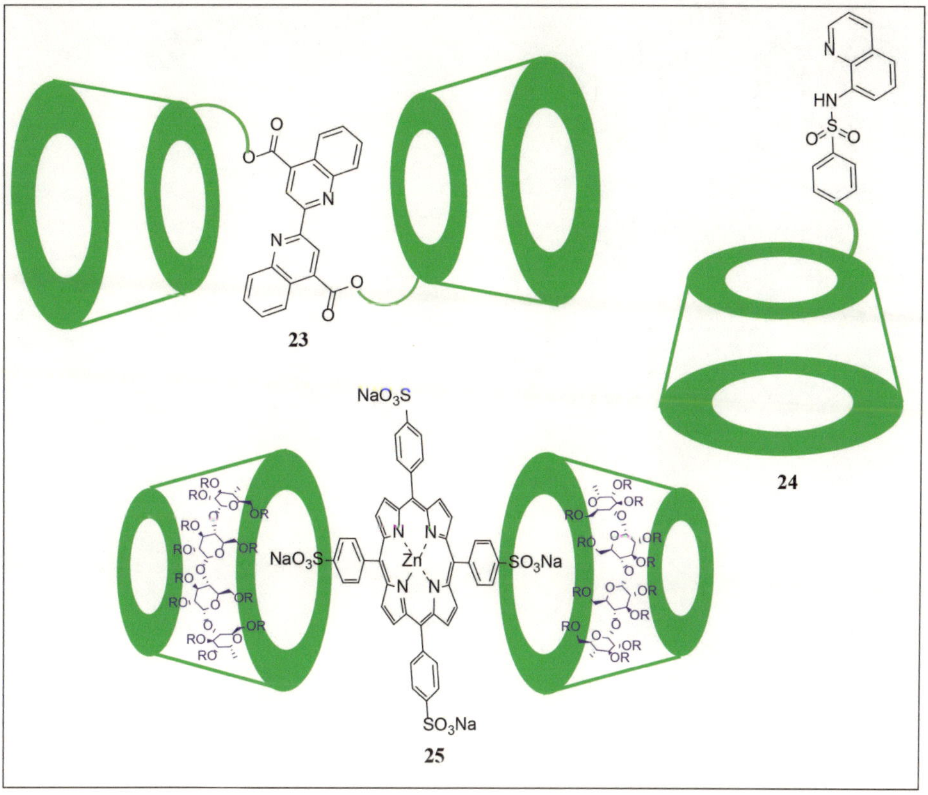

Figure 6.
General structures of β-CD dimer (23) connected by the biquinolino group, quinoline functionalized β-CD sensor (24) for Zn(II) ion and alkylated β-CD/tetraphenylporphyrin-based 2:1 host-guest complex (25).

(II) ion (**Figure 6**). The developed fluorescent sensor (**25**) is composed of alkylated *β*-CD and tetraphenylporphyrin units in the stoichiometric ratio of 2:1. In this case, fluorescent enhancement has been revealed upon the selective complexation of Zn(II) with *meso*-tetraphenylporphyrin among various other interfering metal ions in aqueous medium [38].

4. Cyclodextrin-based supramolecular systems as chemical sensors

Design and construction of the supramolecular architectures utilizing CD units as the key building blocks have attracted an increasing curiosity in the development of chemical sensors [39, 40]. In comparison with CD monomers, the covalently coupled CD-dimers and CD-trimers possess bigger hydrophobic cavities to accommodate the large guest molecules, which make them ideal candidates for chemical sensing. In this context, Ueno and co-workers have reported the *β*-CD dimer fluorescent sensor (**26**) in which the two *β*-CD moieties are linked through a primary face *via* dansyl group and used it to recognize steroids (**Figure 7**) [41]. The fluorescence quenching in the dansyl moiety has been observed upon the inclusion of steroid guest molecules in the hydrophobic cavities of **26**. This is due to the fact that steroid inclusion into hydrophobic cavity brings about the exclusion of dansyl moiety from hydrophobic space to aqueous hydrophilic media. In another event, Reinhoudt's group constructed *β*-CD

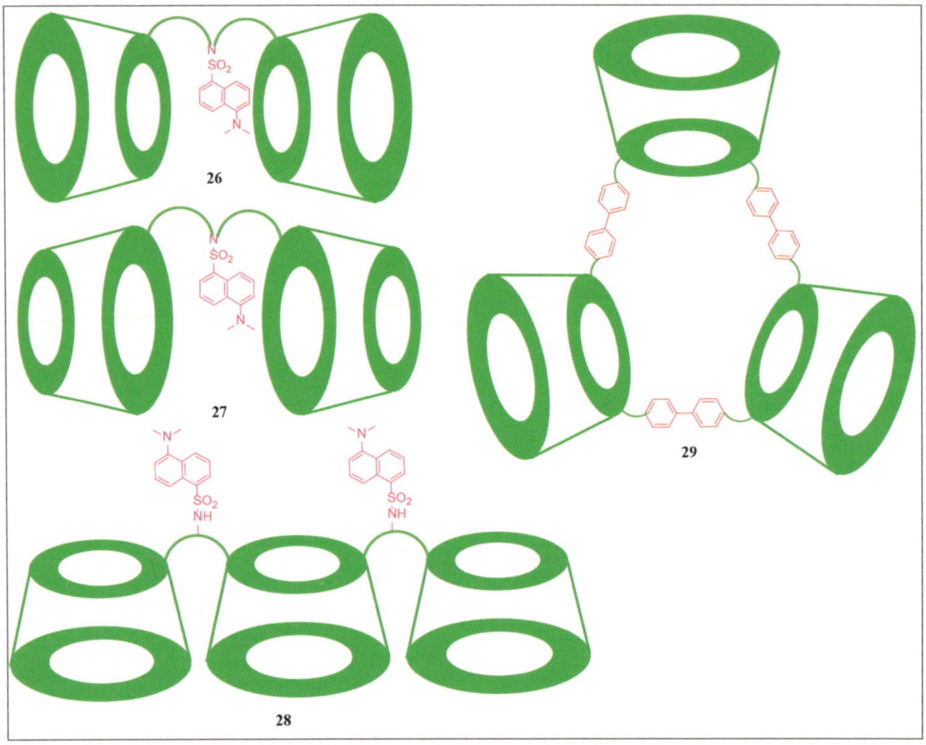

Figure 7.
*Pictorial representation of the fluorescent β-CD-based dimers (**26** and **27**) and trimers (**28** and **29**).*

dimer-based fluorescent sensor (**27**), in which the two *β*-CD subunits are linked through secondary face *via* a dansyl moiety (**Figure 7**) and also revealed different host-guest geometries in comparison with **26** [42]. On the other hand, Kikuchi *et al.* reported the *β*-CD linear trimer-based fluorescent sensor (**28**) consisting of two dansyl moieties as linkers between three *β*-CDs, which in turn signifies the sensing event through host-guest chemistry with bile acids *viz.* cholic acid, lithocholic acid, and deoxycholic acid (**Figure 7**) [43]. On the other hand, Sasaki *et al.* have fabricated permethylated *β*-CD-based fluorescent cyclic trimer (**29**) in which *β*-CD units are bridged through biphenyl moieties (**Figure 7**). From the experimental studies, it was revealed that **29** strongly captures an anthracene derivative possessing two alkyl chains and signifies the binding event *via* fluorescence modulations [44].

Interestingly, sensing conjugates of CDs with macrocyclic hosts employing cooperative molecular recognition phenomenon have also been fabricated by various researchers across the world. In this context, Hayashita and teammates have developed a highly selective hybrid molecular conjugate (**30**) between *γ*-CD and pyrene crown ether [45, 46]. It has been noticed that in the presence of K^+ ion, the emission of pyrene monomer disappears, resulting in excimer emission due to the formation of a 2:1 host-guest sandwiched complex between crown ether and K^+ ion (**Figure 8a**). While, Tong *et al.* have fabricated a conjugate sugar sensing system (**32**) between *β*-CD and pyrene attached boronic acid fluorophore (**Figure 8b**) [47]. Fluorescence enhancement has been noticed upon the sensing of sugar moiety by conjugate system (**32**) as can be inferred from **Figure 8b**. On the other hand, Kaneda *et al.* have constructed a sensing molecular conjugate (**35**) between methylated *α*-CD and crown

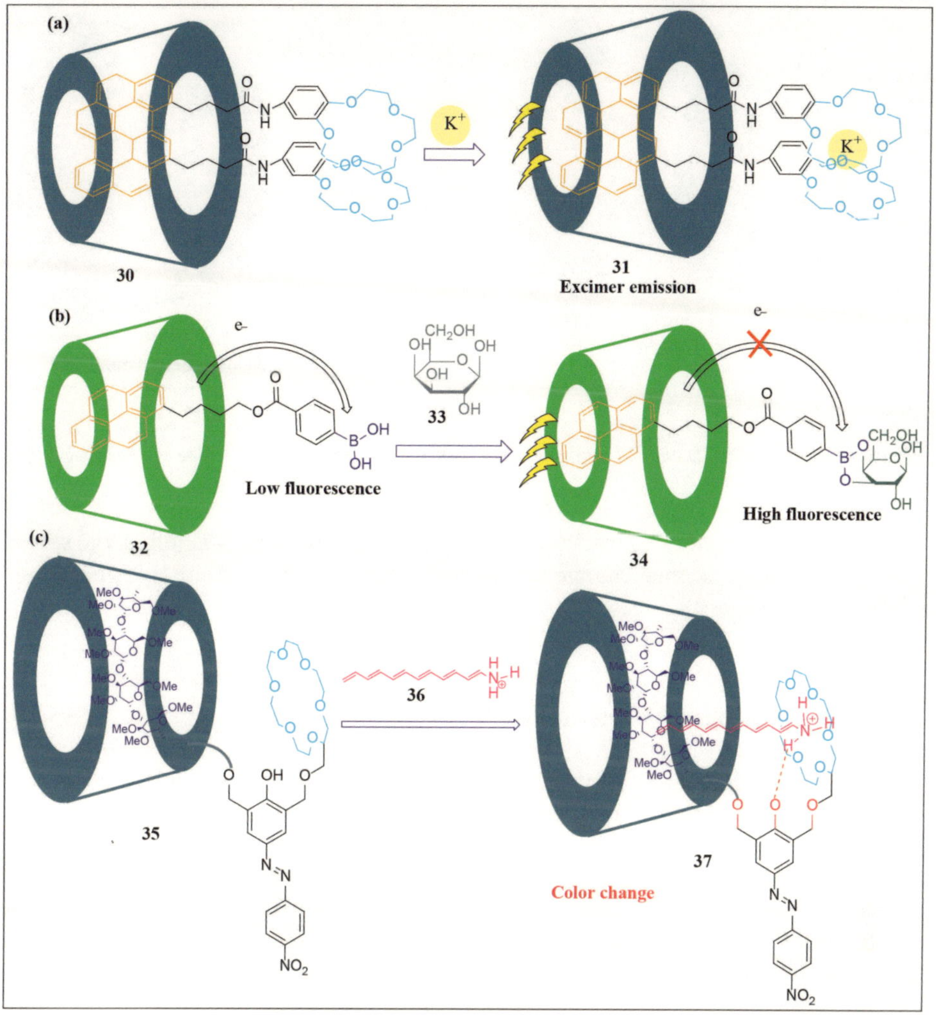

Figure 8.
Schematic representations of CD-based molecular sensing conjugates (a) molecular conjugate (30) of γ-CD with pyrene crown ether, (b) molecular conjugate (32) of β-CD with pyrene-functionalized boronic acid, and (c) molecular conjugate (35) of methylated α-CD with crown ether-functionalized azo-phenyl dye.

ether functionalized azo-phenyl dye (**Figure 8c**) [48, 49]. Interestingly in aqueous media, a prominent color change was noticed upon the addition of 1° or 2° alkylamines to the conjugate sensing system (**35**). However, in aqueous solution, no such color changes were observed upon the addition of 3° alkylamine to **35**. The reason for color change is ascribed to the fact that 1° or 2° alkylamines are strongly bonded to crown ether moiety of **35** and their lipophilic alkyl tails construct a strong complex with the CD framework (**Figure 8c**).

The research group of Anderson has used γ-CD and [2]rotaxane (possessing stilbene axle and terphenylenedicarboxylic acid stoppers) for the preparation of a unique chemosensor **38** (**Figure 9**) [50]. It has been revealed that the stilbene axle of [2] rotaxane offers hydrophobic floor to γ-CD cavity and hence leads to an increase in its affinity to 1000-fold for appropriate guests (**39**) in comparison with simple γ-CD (**3**).

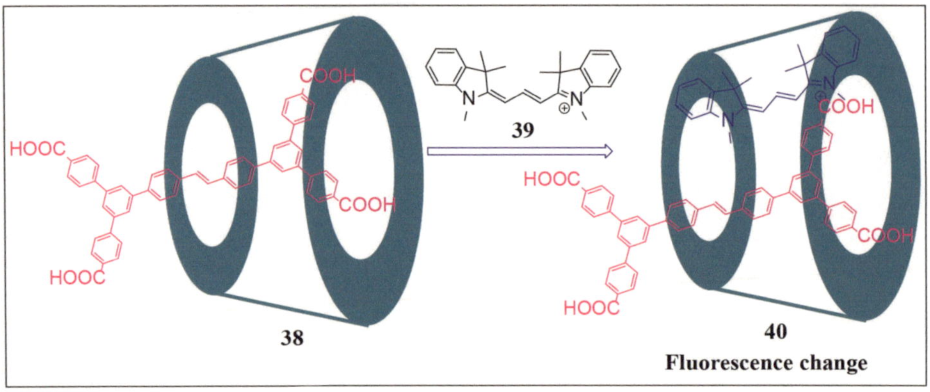

Figure 9.
Schematic illustration of γ-CD and [2] rotaxane-based chemosensor (38) depicting sensing of a guest molecule (39) via *fluorescence change.*

Moreover, stilbene axle of [2]rotaxane also acts as a fluorophore—signifies the sensing event through fluorescence change between chemosensor **38** and suitable guest molecule **39** (**Figure 9**).

5. Cyclodextrin-based electrochemical sensors

Owing to the chief and portable instrumentation, rapid analysis, and high selectivity, as well as sensitivity, in recent years, electrochemical sensing has engrossed a significant courtesy in the recognition of biomolecules and environmentally hazardous pollutants [51–53]. Cyclodextrin-based functional materials have proven to be highly useful in the domain of electrochemical sensing in past decade [54, 55]. These functional materials mainly include CD-based carbon nanomaterials: carbon nanotubes (CNTs), graphene, and conducting polymers. Nowadays, developing the CD-based conducting polymers for the purpose of electrochemical sensing is considered a hot subject of research interest [56]. This is due to the fact that CD-based conducting polymers pasted on electrodes *via* electrooxidation process of monomers offer high stability, good catalytic ability, and electronic features [57]. To this regard, Bouchta *et al.* have fabricated gold electrode with poly(3-methylthiophene)-based γ-CD through electropolymerization process for the electrochemical determination of dopamine, chlorpromazine, 3,4-dihydroxyphenyl alanine, etc. [58]. On the other hand, Luong and co-workers have doped a diamond electrode with boron and sulfobutylether functionalized β-CD along with a composite film of polypyrrole and poly (*N*-acetyltyramine) for the selective electrochemical determination of neurotransmitter dopamine among other interfering analytes, such as ascorbic acid and uric acid [59].

In the context of CD-based carbon nanomaterials, Huang's research group has modified glassy carbon electrode (GCE) by single-walled CNT (SWCNT) and pyrene functionalized β-CD (**42**) in order to determine the 3,3′,4,4′-tetrachlorobiphenyl (**41**) *via* electrochemical impedance method (**Figure 10**). It was noticed by the authors that the pyrene moiety aids in attaching the **42** onto the SWCNT (**43**) sidewall through π-π-stacking interactions, and the guest molecule **41** gets encapsulated by the hydrophobic cavity of **42** [60]. Furthermore, they also reported the electrochemical sensing

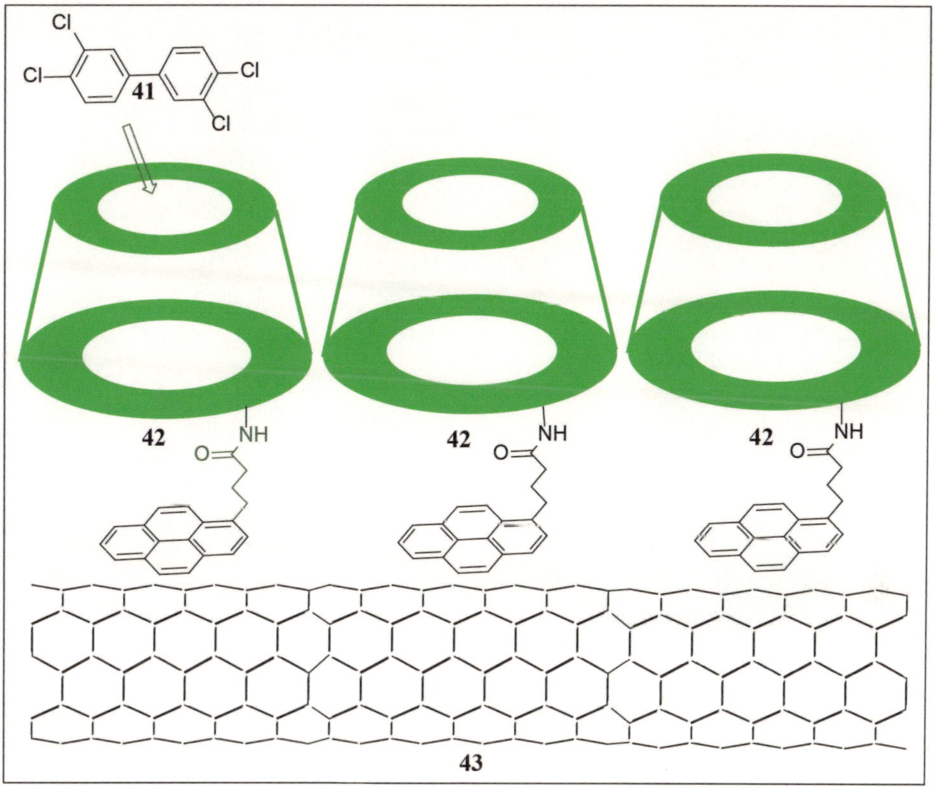

Figure 10.
Schematic illustration of π-π-stacking adsorption of pyrene functionalized β-CD (42) onto the side wall of SWCNT (43) along with the structure of 3,3′,4,4′-tetrachlorobiphenyl (41) guest molecule.

of *p*-nitrophenol using same SWCNT (**43**)-based pyrene functionalized *β*-CD nanohybrids. The nanohybrid traps *p*-nitrophenol in the hydrophobic cavity of **42** with high selectivity and sensitivity with a detection limit of 0.00086 μM [61].

Recognition of chiral enantiomers *via* CD-based electrochemical sensors is of immense importance in the medical and pharmaceutical sciences [62]. Yang and co-workers have recently modified the surface of GCE by hydroxypropyl *β*-CD grafted cellulose, multi-walled CNTs (MWCNTs), and copper ions in order to develop a sensitive electrochemical sensor for the recognition of chiral enantiomers of tryptophan (*D*-Trp/*L*-Trp) [63]. It has been perceived that the fabricated electrochemical sensor has higher affinity toward *L*-Trp in comparison with the *D*-Trp (**Figure 11**). Additionally, the developed electrochemical sensor has been successfully utilized to monitor the quantity of *D*-Trp in racemic mixture. These studies thus pave the way toward the development of realistic chiral platforms for the recognition of diverse chiral molecules. On the other hand, a *β*-CD-based sensitive electrochemical sensor for the recognition of endocrine disrupting agent known as bisphenol A, in an aqueous solution, has been reported through the pasting of MWCNTs (**46**) and graphene oxide (**48**) on screen-printed carbon electrode (SPE) (**Figure 12**) [64]. This versatile system follows a diffusion-controlled mechanism in the sensitive electrochemical sensing of bisphenol A in drinking water with a detection limit of up to 6 nM. These studies thus offer a promising role in the determination of water quality *via* bisphenol A monitoring.

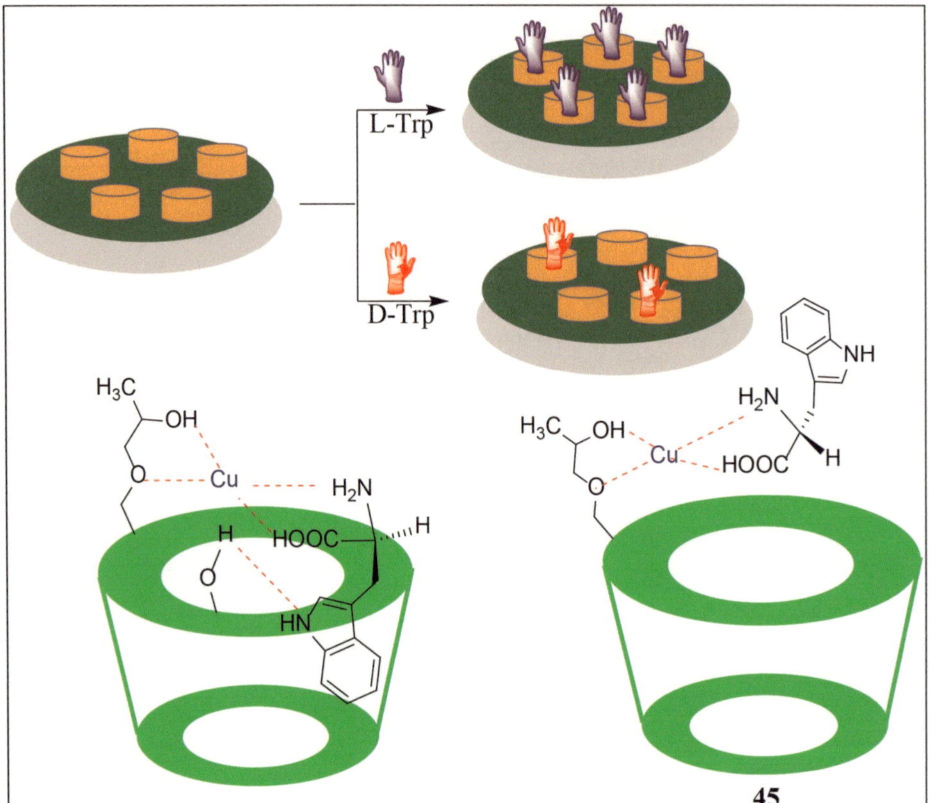

Figure 11.
Schematic illustration of electrochemical recognition of the chiral enantiomers of tryptophan (Trp) via *hydroxypropyl β-CD-based electrochemical sensor.*

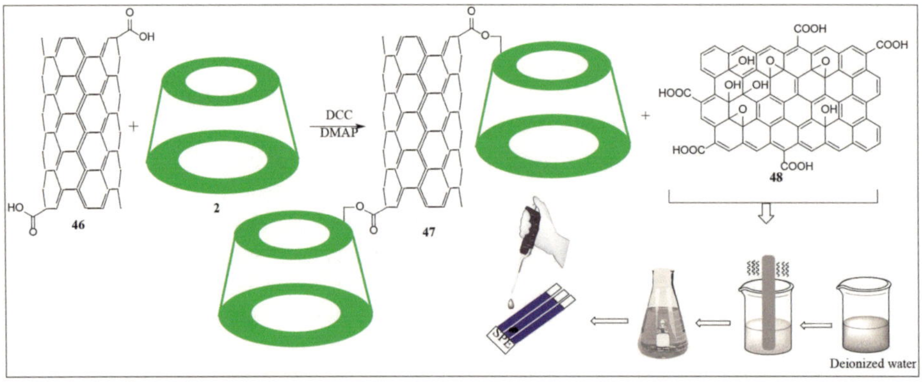

Figure 12.
Schematic illustration for the development of bisphenol A electrochemical sensor via *modification of SPE with β-CD and graphene oxide-functionalized MWCNTs.*

Over the past several years, diverse CD-based potentiometric sensors working through electrochemical means have fruitfully been developed, which find significant applications in medicine, agriculture, environmental monitoring, pharmaceutical sciences, and industries [65, 66]. In this context, Lenik and teammates have developed functionalized β-CDs-based potentiometric sensors for the determination of useful pharmaceutical drugs known as naproxen and ketoprofen. It has been observed that the guest naproxen molecule is partially or completely encapsulated within the cavity of host functionalized β-CD [67, 68]. On the other hand, Amorim *et al.* have also used functionalized β-CDs in the fabrication of potentiometric sensors for psychiatric drug molecules *viz.* diazepam and midazolam [69]. On the other hand, Khaled's research group has fabricated carbon paste electrodes with β-CD based polyvinyl chloride in order to determine diverse acetylcholine derivatives *viz.* butrylcholine, acetylthiocholine, and acetylmethylcholine [70].

6. Cyclodextrin-based polymers as chemical sensors

Due to widespread applicability of π-conjugated polymers in electroluminescence, light-emitting diodes (LEDs), electrical conductivity, and chemical sensors, researchers are curious worldwide to explore fine-tuning of their electrical and optical properties by virtue of stimuli, such as pH, metal ion, and redox reactions [71]. In this context, Harada's research group has constructed β-CD functionalized poly(phenylene ethynylene)-based π-conjugated fluorescent polymer (**49**), which is water soluble and displays blue fluorescence in DMF and green fluorescence in aqueous solutions (**Figure 13**) [72]. Upon the addition of a competitive guest molecule known as 1-adamantanecarboxylic acid (**51**) to **49**-based intermolecular aggregates (**50**), fluorescence color variation from green to blue was observed by the authors (**Figure 13**). This can be ascribed to the fact that 1-adamantane-carboxylic acid (**51**) complexation with β-CD units of **49** results in the dissociation of various intermolecular π-stacking interactions of polymeric backbone. In fact, the repulsive interactions between the anionic counterparts of 1-adamantane-carboxylic acid (**51**) hinder the polymeric chains to come into the aggregation. By adding electron-accepting adamantane-functionalized viologen derivative (**53**) to the π-conjugated fluorescent polymer (**49**), large fluorescence quenching was seen due to the formation of inclusion complex between β-CD moiety of polymer (**49**) and adamantane group of viologen derivative (**53**). Further, the host-guest interaction, assisted in upholding viologens on the polymeric side chain, results in an adept electron transfer between polymeric backbone to viologen unit of **54** (**Figure 14**).

As coumarin and pyrene scaffolds are of great importance, owing to their vital role in biological systems and sensing arena as well [73]. In this regard, Ueno's group has synthesized β-CD-peptide hybrid polymeric conjugate (**55**), having pyrene as a donor moiety and coumarin as an acceptor one (**Figure 15**) [74, 75]. The coumarin moiety is encapsulated within the hydrophobic cavity of β-CD and thus offers strong fluorescence to the hybrid polymeric conjugate (**55**) *via* fluorescence resonance energy transfer (FRET) from donor pyrene unit to acceptor coumarin unit ("FRET-ON" response). It has been remarked that the addition of competitive guest molecule, namely hyodeoxycholic acid (**56**), leads to the decrease in fluorescence by virtue of the exclusion of coumarin moiety from inside of the β-CD hydrophobic cavity to outside. This, in turn, leads to the association between coumarin and pyrene units and offers the "FRET-OFF" response (**Figure 15**) [74]. Inouye and Fujimoto have developed methylated β-CD-DNA hybrid polymeric conjugate (**58**) sensor for porphyrin derivatives (**Figure 16**) [76]. They observed that **58** captures

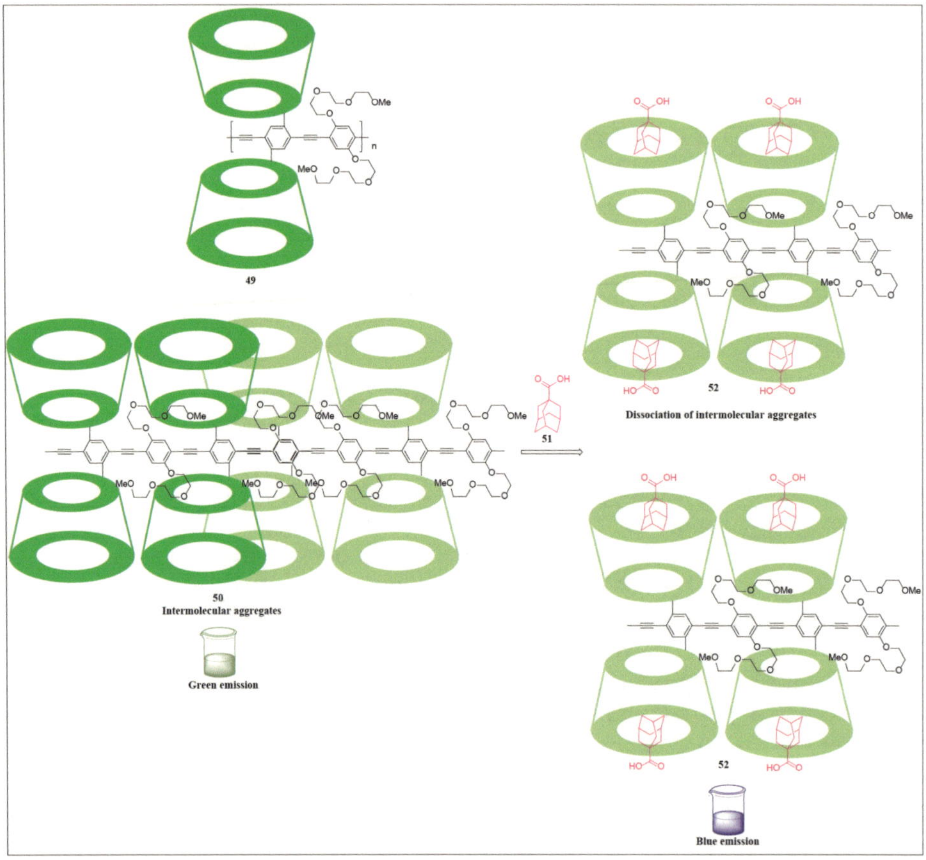

Figure 13.
Schematic illustration of fluorescence color variation from green to blue upon addition of 1-adamantanecarboxylic acid (51) to the intermolecular aggregates (50).

meso-tetraphenylporphyrin sulfonate (**59**) in a 2:1 stoichiometric ratio. This, in turn, induces the formation of DNA duplex structure and results in excimer emission (**Figure 16**).

On the other hand, quite recently, Badiei and co-workers have established a *β*-CD-based cross-linked polymer, the CD-nanosponge (**62**) with pyromellitic anhydride (**61**) cross-linker for the selective and sensitive detection of diclofenac among various other interfering analytes, such as ibuprofen, morphine, amphetamine, and codeine (**Figure 17**) [77]. For diclofenac, they have observed a detection limit of 0.92 μM and linear range of 1–33 μM. Interestingly for real-world applications, the established *β*-CD-based fluorescence probe has the utility to determine the concentration of diclofenac in commercially accessible pharmaceutical tablets.

7. Cyclodextrin-nanocarbon hybrids as chemical sensors

Over the past several decades, carbon nanomaterials for instance carbon nanotubes, fullerene and nanodiamonds, and other polyaromatic hydrocarbons because their unique electrical, structural, and mechanical properties have riveted a significant interest of the researchers across the globe to meet the challenge of

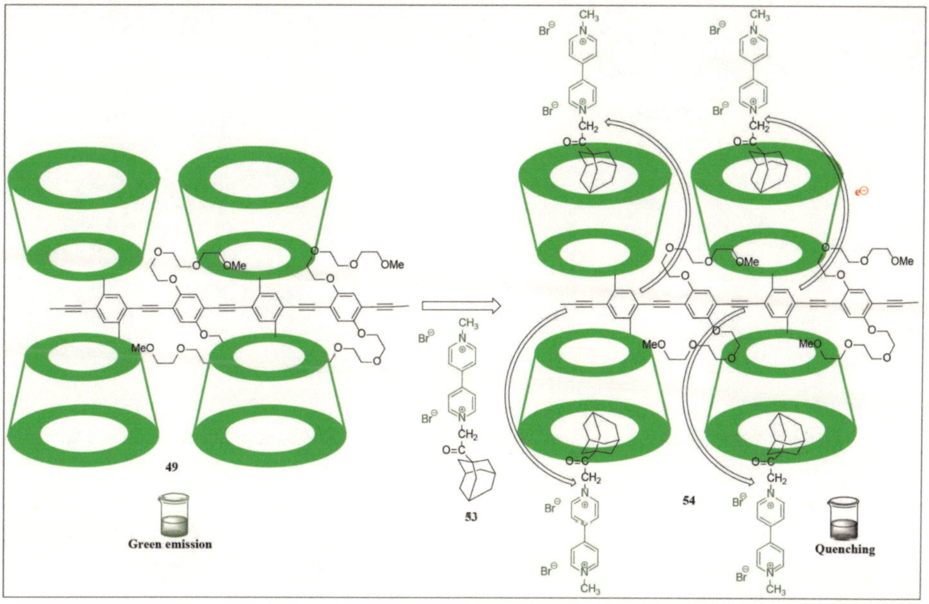

Figure 14.
Schematic depiction of fluorescence quenching upon the addition of adamantane functionalized viologen derivative (53) to π-conjugated fluorescent polymer (49).

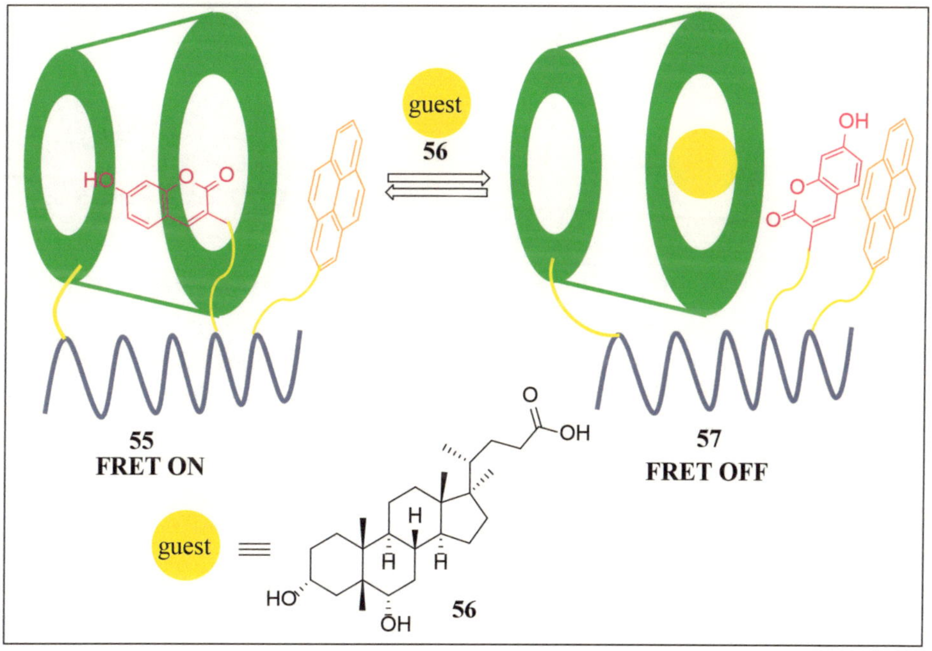

Figure 15.
Schematic representation of hyodeoxycholic acid (56) assisted structural variation of β-CD-peptide hybrid polymeric conjugate (55).

constructing the CD-based sensors through hybridization with carbon nanomaterials [78–81]. In this context, Fujita and Yuan *et al.* have reported a β-CD-fullerene hybrid conjugate (**63**), which has the ability to quench the fluorescence of rhodamine B (**64**),

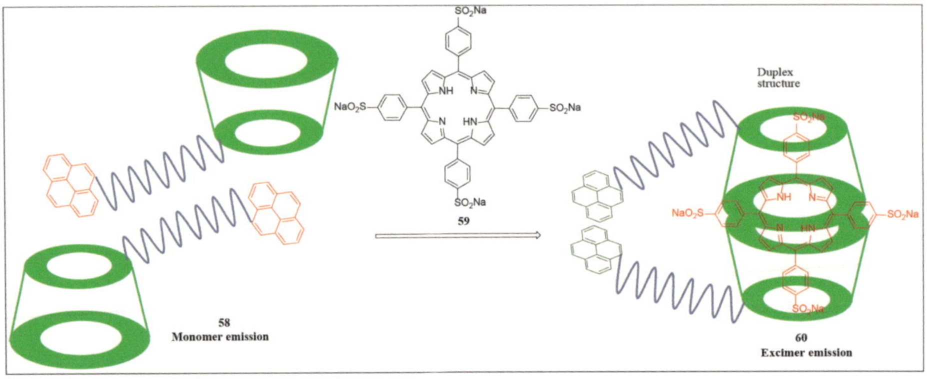

Figure 16.
Schematic representation of guest (59) assisted structural variation of methylated β-CD-DNA hybrid fluorescent polymeric conjugate (58).

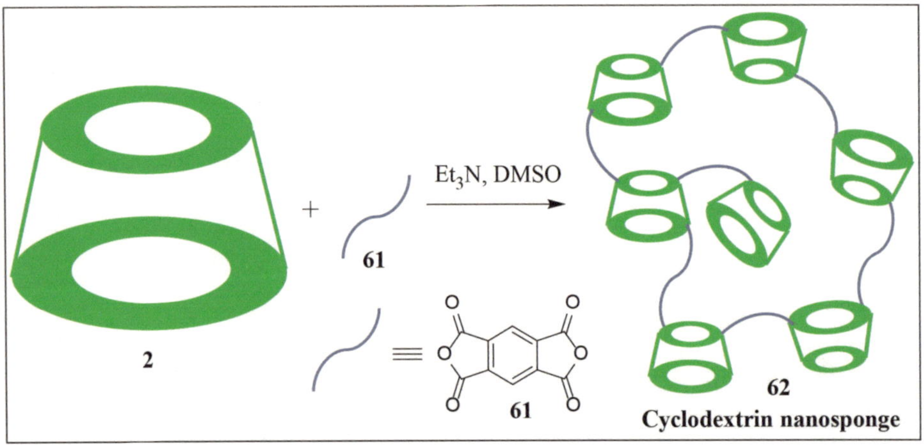

Figure 17.
Schematic illustration of β-CD based cross-linked polymer (62) with pyromellitic anhydride (61) acting as cross-linker.

after its capture in the hydrophobic cavity of *β*-CD (**Figure 18**) [82]. On the other hand, Harada and co-workers have constructed pyrene-functionalized *β*-CD-SWCNT hybrid conjugate (**67**) and utilized it in the development of stimuli-responsive supramolecular hydrogel (**Figure 19**). It has been observed that the addition of sodium adamantane carboxylate as a competitive guest leads toward the conversion of a gel to sol [83]. Remarkably, the Stoddart's group has decorated pyrene-functionalized *β*-CDs on SWCNT hybridized with field-effect transistors (FETs) in order to sense typical organic guest molecules *viz*. 1-adamantanol, sodium cholate, 1-adamantane carboxylic acid. It was observed that the FET characteristics of these hybrid nano-conjugates are highly sensitive and dependent on the association constants between *β*-CDs and competitive organic guest molecules [84].

8. Cyclodextrin-nanoparticle hybrids as chemical sensors

In recent years, scientific community has exploited the unique property of gold nanoparticle aggregation in the design and construction of various optical sensory

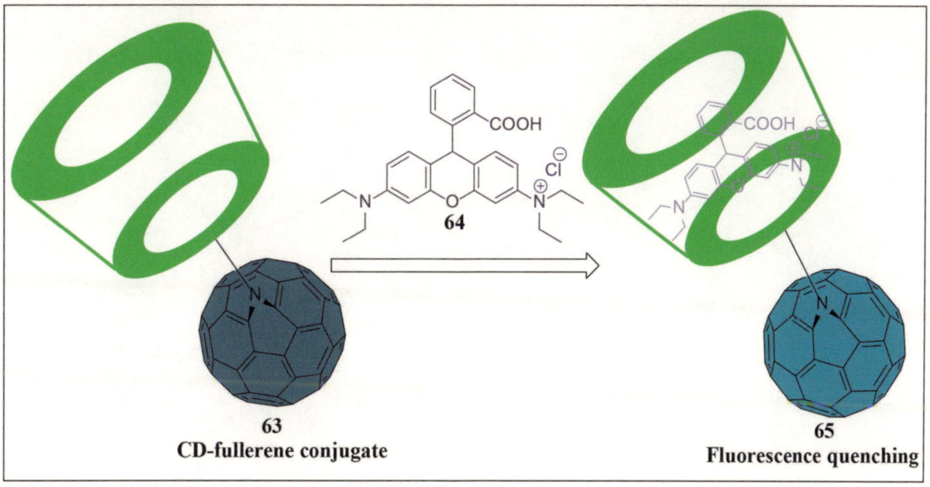

Figure 18.
Schematic illustration of fluorescence quenching of rhodamine B (64) via the β-CD-fullerene hybrid conjugate (63).

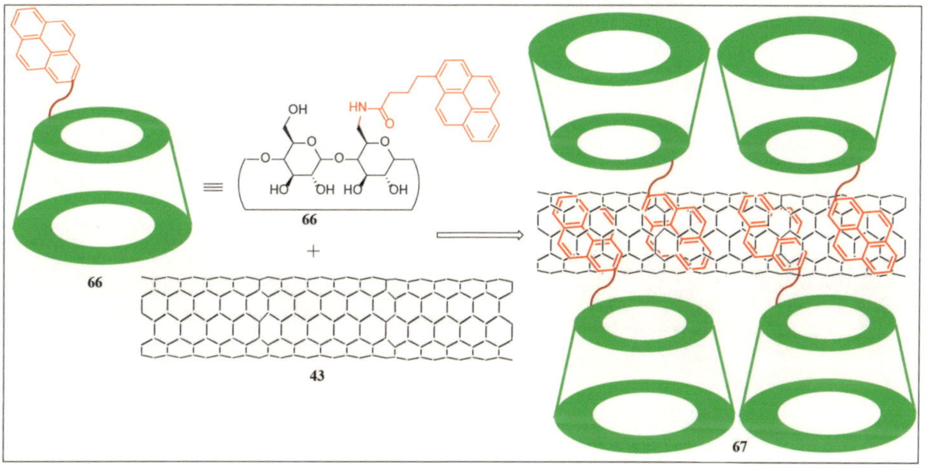

Figure 19.
Schematic depiction of the formation of pyrene functionalized β-CD-SWCNT hybrid conjugate (67).

devices where the sensing mechanism is perceived through color variation from red to purple/blue. In this regard, Kaifer and teammates have constructed β-CD functionalized gold nanoparticles (**68**) and noticed that the addition of ferrocene dimer (**69**) as a competitive guest to colloidal solutions of **68** primarily results in a red shift followed by the precipitation of a red solid (**Figure 20**) [85, 86]. These observations were not perceived in case of addition of ferrocene in methanol. This indicates that the ferrocene dimer operates as a linker between diverse gold nanoparticles, and hence helps in their aggregation. This aggregation in turn offers a color change, which signifies the sensing event with typical guest molecules. The same group also utilized the γ-CD in combination with gold nanoparticles for the sensing of well-known carbon nanomaterial known as fullerene (C_{60}) through aggregation phenomenon [87]. In another event, Tang *et al.* constructed a highly selective as well as sensitive glucose

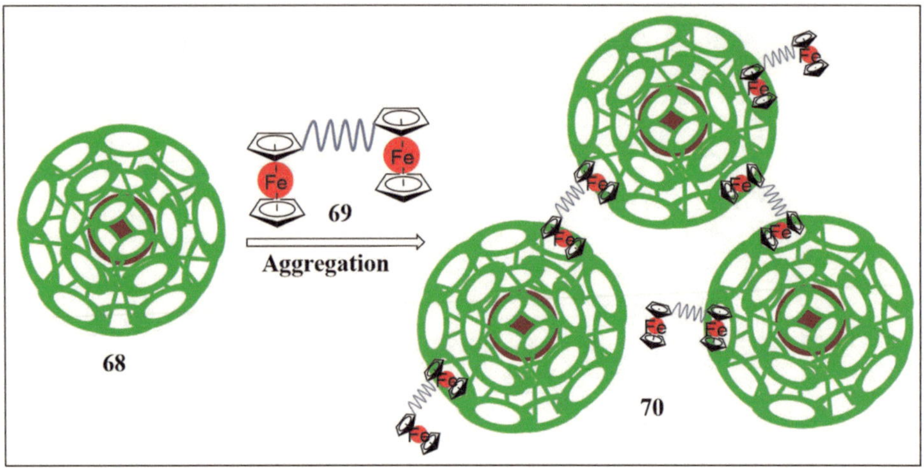

Figure 20.
Schematic view of β-CD functionalized gold nanoparticles (68) aggregation upon addition of ferrocene dimer (69).

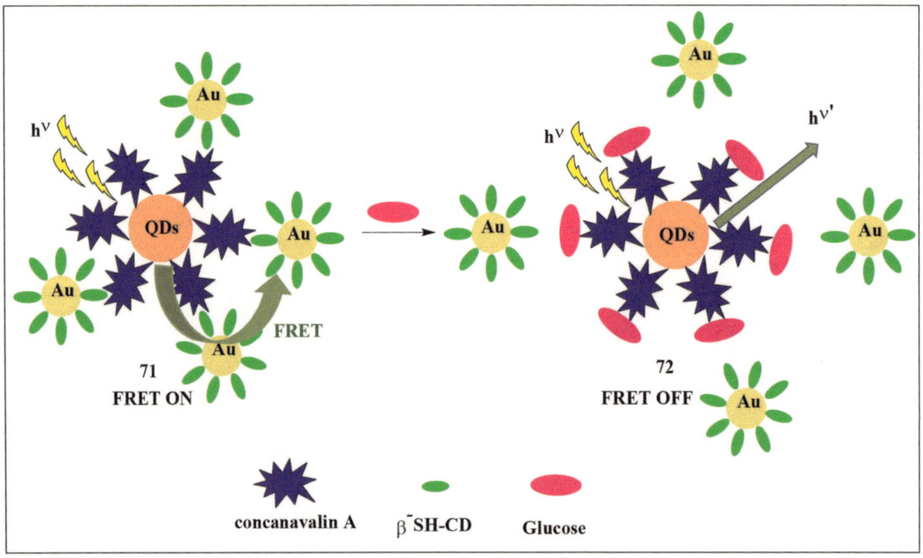

Figure 21.
Schematic illustration of FRET on/off mechanism operating in nanobiosensor (71) constructed from concanavalin A fabricated CdTe (QDs) and thiolated β-CD functionalized AuNps.

nanobiosensor (**71**) in serum, which operates through FRET between concanavalin A fabricated CdTe quantum dots (QDs; energy donors) and thiolated *β*-CD functionalized gold nanoparticles (AuNps) acting as energy acceptors (**Figure 21**) [88]. Quite recently, Bindu and co-workers have functionalized gold nanoparticles by the *β*-CD in order to detect heavy metal ions in aquatic realm. Captivatingly, the sensitivity of the developed *β*-CD-gold nanoparticle hybrid conjugate toward copper was found to be 1.788 mM [89]. Elgamouz and teammates have functionalized silver nanoparticles by *β*-CD to develop a nanoprobe having the ability to encapsulate creatinine through the colorimetric response. The established *β*-CD-silver

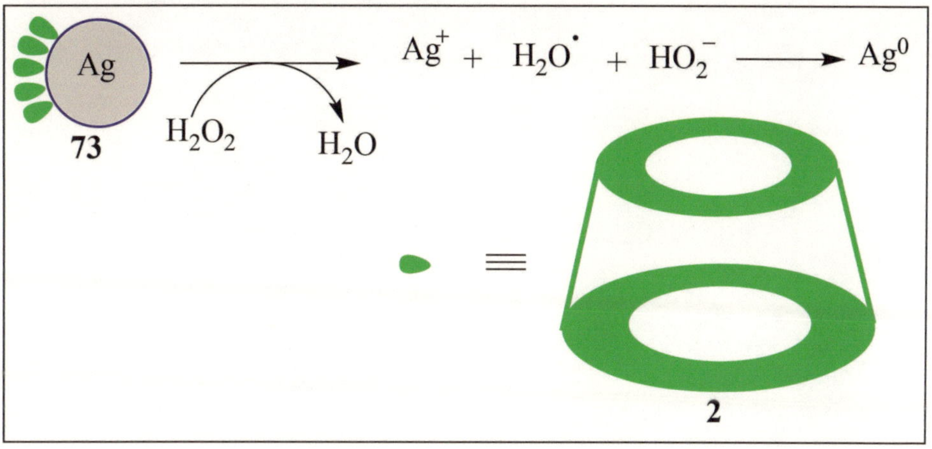

Figure 22.
Schematic representation of the detection of H_2O_2 via β-CD-AgNP-based nano colorimetric probe (73).

nanoparticle (AgNP)-based nano-colorimetric probe (**73**) has successfully been applied in the detection of reactive oxygen species *viz.* H_2O_2 in human urine samples by these authors (**Figure 22**) [90].

9. Conclusions and outlook

In summary, this chapter discusses the conceptual background as well as evolutionary developments of chemical sensing in cyclodextrin-based monomers, dimers, clusters, and nano-assemblies with a detection limit up to μM/nM level. The sensing event of various guest molecules *via* optical and electrochemical signatures on CD-based sensors endows them characteristics and features, which have been elaborated. In fact, the domain of CD-based chemical sensors has established its firm ground in supramolecular chemistry, biochemistry, polymer chemistry, pharmaceutical chemistry, and nanotechnology. Utilizing the fundamental principles and concepts of chemistry in combination with CD-based chemistry, diverse novel chemical sensors having high sensitivity, high functionality, and wide versatility have been constructed for analytes of typical interest. The authors are of the view that this chapter will offer new dimensions to CD-based chemical sensors and may guide the readers to develop a better understanding of cyclodextrins.

Acknowledgements

We are highly thankful to DST-SERB New Delhi for financial support (Project File no. ECR/2017/000821). I. A. R. thanks CSIR, New Delhi, for the award of the JRF and SRF. A.H. thanks UGC for non-NET fellowship. R. A. thanks Jamia Millia Islamia, New Delhi, for providing the necessary research facilities.

Conflicts of interest

The authors declare no conflicts of interest.

Author details

Ishfaq Ahmad Rather, Ahmad Hasan and Rashid Ali*
Department of Chemistry, Organic and Supramolecular Functional Materials Research Laboratory, Jamia Millia Islamia, Okhla, New Delhi, India

*Address all correspondence to: rali1@jmi.ac.in

References

[1] Rather IA, Wagay SA, Ali R. Emergence of anion-π interactions: The land of opportunity in supramolecular chemistry and beyond. Coordination Chemistry Reviews. 2020;**415**:213327. DOI: 10.1016/j.ccr.2020.213327

[2] Rather IA, Ali R. Indicator displacement assays: From concept to recent developments. Organic & Biomolecular Chemistry. 2021;**19**: 5926-5981. DOI: 10.1039/D1OB00518A

[3] Wagay SA, Rather IA, Ali R. Functionalized calix[4]pyrroles: Emerging class of ion-pair receptors in supramolecular chemistry. Materials Today: Proceedings. 2021;**36**: 657-678. DOI: 10.1016/j.matpr.2020.04.648

[4] Rather IA, Ali R. A catalytic and solvent-free approach for the synthesis of diverse functionalized dipyrromethanes (DPMs) and calix[4] pyrroles (C4Ps). Green Chemistry. 2021; **23**:5849-5855. DOI: 10.1039/D1GC01515J

[5] Rather IA, Wagay SA, Hasnain MS, Ali R. New dimensions in calix[4] pyrrole: The land of opportunity in supramolecular chemistry. RSC Advances. 2019;**9**:38309-38344. DOI: 10.1039/C9RA07399J

[6] Lee JY, Root HD, Ali R, An W, Lynch VM, Bähring S, et al. Ratiometric turn-on fluorophore displacement ensembles for nitroaromatic explosives detection. Journal of the American Chemical Society. 2020;**142**:19579-19587. DOI: 10.1021/jacs.0c08106

[7] Sinn S, Biedermann F. Chemical sensors Based on cucurbit[n]uril macrocycles. Israel Journal of Chemistry. 2018;**58**:357-412. DOI: 10.1002/ijch.201700118

[8] Ogoshi T, Harada A. Chemical sensors based on cyclodextrin derivatives. Sensors. 2008;**8**:4961-4982. DOI: 10.3390/s8084961

[9] Hapiot F, Tilloy S, Monflier E. Cyclodextrins as supramolecular hosts for organometallic complexes. Chemical Reviews. 2006;**106**:767-781. DOI: 10.1021/cr050576c

[10] Votava M, Ravoo BJ. Principles and applications of cyclodextrin liquid crystals. Chemical Society Reviews. 2021;**50**:10009-10024. DOI: 10.1039/D0CS01324B

[11] Tian B, Liu Y, Liu J. Smart stimuli-responsive drug delivery systems based on cyclodextrin: A review. Carbohydrate Polymers. 2021;**251**:116871. DOI: 10.1016/j.carbpol.2020.116871

[12] Zhao X, Wang Y, Zhang P, Lu Z, Xiao Y. Recent advances of molecularly imprinted polymers based on cyclodextrin. Macromolecular Rapid Communications. 2021;**42**:2100004. DOI: 10.1002/marc.202100004

[13] Connors KA. The stability of cyclodextrin complexes in solution. Chemical Reviews. 1997;**97**:1325-1358. DOI: 10.1021/cr960371r

[14] Rekharsky MV, Inoue Y. Complexation thermodynamics of cyclodextrins. Chemical Reviews. 1998; **98**:1875-1918. DOI: 10.1021/cr970015o

[15] Khan AR, Forgo P, Stine KJ, D'Souza VT. Methods for selective modifications of cyclodextrins. Chemical Reviews. 1998;**98**:1977-1996. DOI: 10.1021/cr970012b

[16] Thomas SW, Joly GD, Swager TM. Chemical sensors based on amplifying

fluorescent conjugated polymers. Chemical Reviews. 2007;**107**:1339-1386. DOI: 10.1021/cr0501339

[17] Giannetti A, Bocková M. Optical chemosensors and biosensors. Chem. 2020;**8**:33. DOI: 10.3390/chemosensors8020033

[18] Krämer J, Kang R, Grimm LM, De Cola L, Picchetti P, Biedermann F. Molecular probes, Chemosensors, and nanosensors for optical detection of biorelevant molecules and ions in aqueous media and biofluids. Chemical Reviews. 2022;**122**:3459-3636. DOI: 10.1021/acs.chemrev.1c00746

[19] Chadha R, Das A, Lobo J, Meenu VO, Paul A, Ballal A, et al. γ-Cyclodextrin capped silver and gold nanoparticles as colorimetric and Raman sensor for detecting traces of pesticide "Chlorpyrifos" in fruits and vegetables. Colloids and Surfaces A: Physicochemical and Engineering Aspects. 2022;**641**:128558. DOI: 10.1016/j.colsurfa.2022.128558

[20] Yu X, Wan S, Wu W, Yang C, Lu W. γ-Cyclodextrin-based [2]rotaxane stoppered with gold(I)–ethynyl complexation: Phosphorescent sensing for nitroaromatics. Chemical Communications. 2022;**58**:6284-6287. DOI: 10.1039/D2CC02256G

[21] Ikeda H, Murayama T, Ueno A. Skeleton-selective fluorescent chemosensor based on cyclodextrin bearing a 4-amino-7-nitrobenz-2-oxa-1,3-diazole moiety. Organic & Biomolecular Chemistry. 2005;**3**:4262. DOI: 10.1039/b508477f

[22] Liu Y, Shi J, Guo D-S. Novel permethylated β-cyclodextrin derivatives appended with chromophores as efficient fluorescent sensors for the molecular recognition of bile salts. The Journal of Organic Chemistry. 2007;**72**:8227-8234. DOI: 10.1021/jo071131m

[23] Kuwabara T, Nakamura A, Ueno A, Toda F. Inclusion complexes and guest-induced color changes of pH-indicator-modified.Beta.-cyclodextrins. The Journal of Physical Chemistry. 1994;**98**: 6297-6303. DOI: 10.1021/j100076a011

[24] Ueno A, Kuwabara T, Nakamura A, Toda F. A modified cyclodextrin as a guest responsive colour-change indicator. Nature. 1992;**356**:136-137. DOI: 10.1038/356136a0

[25] Barcza L, Buvári-Barcza Á. Complex formation of cyclomalto-octaose with tetrabromophenolphthalein and some related compounds. Carbohydrate Research. 1989;**192**:103-109. DOI: 10.1016/0008-6215(89)85170-5

[26] Kuwabara T, Nakamura A, Ueno A, Toda F. Supramolecular thermochromism of a dye-appended β-cyclodextrin. Journal of the Chemical Society, Chemical Communications. 1994:689-690. DOI: 10.1039/C39940000689

[27] Aoyagi T, Nakamura A, Ikeda H, Ikeda T, Mihara H, Ueno A. Alizarin yellow-modified β-cyclodextrin as a guest-responsive absorption change sensor. Analytical Chemistry. 1997;**69**: 659-663. DOI: 10.1021/ac960727z

[28] Joseph V, Warhaftig O, Klein S, Levine M. Paper-based manganese and β-cyclodextrin sensors for colorimetric sulfur dioxide detection. Analytica Chimica Acta. 2022;**1200**:339629. DOI: 10.1016/j.aca.2022.339629

[29] Haider JM, Pikramenou Z. Photoactive metallocyclodextrins: Sophisticated supramolecular arrays for the construction of light activated

miniature devices. Chemical Society Reviews. 2005;**34**:120. DOI: 10.1039/b203904b

[30] Kameda T, Takaizumi M, Kumagai S, Saito Y, Yoshioka T. Preparation of Zn–Al layered double hydroxide intercalated with carboxymethyl-*β*-cyclodextrin by anion exchange method and its Ni^{2+} adsorption property. Soft Materials. 2021;**19**: 139-147. DOI: 10.1080/1539445X.2020.1779088

[31] Pikramenou Z, Yu J, Lessard RB, Ponce A, Wong PA, Nocera DG. Luminescence from supramolecules triggered by the molecular recognition of substrates. Coordination Chemistry Reviews. 1994;**132**:181-194. DOI: 10.1016/0010-8545(94)80039-1

[32] Mortellaro MA, Nocera DG. A supramolecular chemosensor for aromatic hydrocarbons. Journal of the American Chemical Society. 1996;**118**: 7414-7415. DOI: 10.1021/ja961323r

[33] Michels JJ, Huskens J, Reinhoudt DN. Noncovalent binding of sensitizers for lanthanide(iii) luminescence in an EDTA-bis(*β*-cyclodextrin) ligand. Journal of the American Chemical Society. 2002; **124**:2056-2064. DOI: 10.1021/ja017025y

[34] Haider JM, Pikramenou Z. Metal assembly of cyclodextrin recognition sites. European Journal of Inorganic Chemistry. 2001:189-194. DOI: 10.1002/1099-0682(20011)2001:1<189::AID-EJIC189>3.0.CO;2-N

[35] Heck R, Dumarcay F, Marsura A. New scaffolds for supramolecular chemistry: Upper-rim fully tethered 5-methyleneureido-5′-methyl-2,2′-bipyridyl cyclodextrins. Chemistry–A European Journal. 2002:8 2438. DOI: 10.1002/1521-3765(20020603)8:11<2438::AID-CHEM2438>3.0.CO;2-A

[36] Liu Y, Song Y, Chen Y, Li X-Q, Ding F, Zhong R-Q. Biquinolino-modified *β* -cyclodextrin dimers and their metal complexes as efficient fluorescent sensors for the molecular recognition of steroids. Chemistry–A European Journal. 2004;**10**:3685-3696. DOI: 10.1002/chem.200305724

[37] Liu Y, Zhang N, Chen Y, Wang L-H. Fluorescence sensing and binding behavior of aminobenzenesulfonamido-quinolino-*β*-cyclodextrin to Zn^{2+}. Organic Letters. 2007;**9**:315-318. DOI: 10.1021/ol062816w

[38] Yang R, Li K, Wang K, Zhao F, Li N, Liu F. Porphyrin assembly on *β*-cyclodextrin for selective sensing and detection of a zinc ion based on the dual emission fluorescence ratio. Analytical Chemistry. 2003;**75**:612-621. DOI: 10.1021/ac020467n

[39] Philip S, Kuriakose S, Mathew T. Designing of a *β*-Cyclodextrin -based supramolecular fluorescent sensor doped with superparamagnetic α-Fe_2O_3 nanoparticles with improved light fastness, thermal, and photoluminescent properties. Journal of Applied Polymer Science. 2021;**138**:51004. DOI: 10.1002/app.51004

[40] Singh VR, Pandey SP, Singh PK. A polyelectrolyte based supramolecular assembly for ratiometric sensing of ATP with very high discrimination from pyrophosphate. Journal of Molecular Liquids. 2021;**328**:115314. DOI: 10.1016/j.molliq.2021.115314

[41] Nakamura M, Ikeda T, Nakamura A, Ikeda H, Ueno A, Toda F. Remarkable molecular recognition of dansyl-modified cyclodextrin dimer. Chemistry

Letters. 1995;**24**:343-344. DOI: 10.1246/cl.1995.343

[42] de Jong MR, Engbersen JFJ, Huskens J, Reinhoudt DN. Cyclodextrin dimers as receptor molecules for steroid sensors. Chemistry–A European Journal. 2000;**6**:4034-4040. DOI: 10.1002/1521-3765(20001103)6:21<4034::AID-CHEM4034>3.0.CO;2-3

[43] Kikuchi T, Narita M, Hamada F. Synthesis of bis dansyl-modified *β*-cyclodextrin liner trimer having multi-recognition sites and high hydrophobic environment. Tetrahedron. 2001;**57**: 9317-9324. DOI: 10.1016/S0040-4020(01)00935-8

[44] Sasaki K, Nagasaka M, Kuroda Y. New cyclodextrin dimer and trimer: Formation of biphenyl excimer and their molecular recognition. Chemical Communications. 2001:2630-2631. DOI: 10.1039/b107987e

[45] Yamauchi A, Hayashita T, Kato A, Nishizawa S, Watanabe M, Teramae N. Selective potassium ion recognition by benzo-15-crown-5 fluoroionophore/*γ*-cyclodextrin complex sensors in water. Analytical Chemistry. 2000;**72**: 5841-5846. DOI: 10.1021/ac000741i

[46] Hayashita T, Qing D, Minagawa M, Lee JC, Ku CH, Teramae N. Highly selective recognition of lead ion in water by a podand fluoroionophore/*γ*-cyclodextrin complex sensor. Chemical Communications. 2003:2160-2161. DOI: 10.1039/B305758E

[47] Tong A-J, Yamauchi A, Hayashita T, Zhang Z-Y, Smith BD, Teramae N. Boronic acid fluorophore/*β*-cyclodextrin complex sensors for selective sugar recognition in water. Analytical Chemistry. 2001;**73**:1530-1536. DOI: 10.1021/ac001363k

[48] Jung JH, Lee SJ, Kim JS, Lee WS, Sakata Y, Kaneda T. *α*-CD/crown-appended diazophenol for selective sensing of amines. Organic Letters. 2006;**8**:3009-3012. DOI: 10.1021/ol060923k

[49] Jung JH, Lee HY, Jung SH, Lee SJ, Sakata Y, Kaneda T. A color version of the Hinsberg test: Permethylated cyclodextrin and crown-appended azophenol for highly selective sensing of amines. Tetrahedron. 2008;**64**: 6705-6710. DOI: 10.1016/j.tet.2008.05.013

[50] Klotz EJF, Claridge TDW, Anderson HL. Homo- and hetero-[3]rotaxanes with two *π*-systems clasped in a single macrocycle. Journal of the American Chemical Society. 2006;**128**: 15374-15375. DOI: 10.1021/ja0665139

[51] Simoska O, Gaffney EM, Minteer SD, Franzetti A, Cristiani P, Grattieri M, et al. Recent trends and advances in microbial electrochemical sensing technologies: An overview. Current Opinion in Electrochemistry. 2021;**30**:100762. DOI: 10.1016/j.coelec.2021.100762

[52] Lu S-M, Peng Y-Y, Ying Y-L, Long Y-T. Electrochemical sensing at a confined space. Analytical Chemistry. 2020;**92**:5621-5644. DOI: 10.1021/acs.analchem.0c00931

[53] Rather IA, Sofi FA, Bhat MA, Ali R. Synthesis of novel one-walled meso - phenylboronic acid-functionalized calix[4]pyrrole: A highly sensitive electrochemical sensor for dopamine. ACS Omega. 2022;**7**:15082-15089. DOI: 10.1021/acsomega.2c00926

[54] Zhu G, Yi Y, Chen J. Recent advances for cyclodextrin-based materials in electrochemical sensing. TrAC Trends in Analytical Chemistry.

2016;**80**:232-241. DOI: 10.1016/j.trac.2016.03.022

[55] Healy B, Yu T, da Silva Alves DC, Okeke C, Breslin CB. Cyclodextrins as supramolecular recognition systems: Applications in the fabrication of electrochemical sensors. Materials. 2021; **14**:1668. DOI: 10.3390/ma14071668

[56] Bae J, Shin K, Kwon OS, Hwang Y, An J, Jang A, et al. A succinct review of refined chemical sensor systems based on conducting polymer–cyclodextrin hybrids. Journal of Industrial and Engineering Chemistry. 2019;**79**:19-28. DOI: 10.1016/j.jiec.2019.06.051

[57] Fritea L, Le Goff A, Putaux J-L, Tertis M, Cristea C, Săndulescu R, et al. Design of a reduced-graphene-oxide composite electrode from an electropolymerizable graphene aqueous dispersion using a cyclodextrin-pyrrole monomer. Application to dopamine biosensing. Electrochimica Acta. 2015; **178**:108-112. DOI: 10.1016/j.electacta.2015.07.124

[58] Bouchta D, Izaoumen N, Zejli H, El Kaoutit M, Temsamani KR. A novel electrochemical synthesis of poly-3-methylthiophene-γ-cyclodextrin film☆application for the analysis of chlorpromazine and some neurotransmitters. Biosensors & Bioelectronics. 2005;**20**:2228-2235. DOI: 10.1016/j.bios.2004.12.004

[59] Shang F, Zhou L, Mahmoud KA, Hrapovic S, Liu Y, Moynihan HA, et al. Selective nanomolar detection of dopamine using a boron-doped diamond electrode modified with an electropolymerized sulfobutylether-β-cyclodextrin-doped poly(N -acetyltyramine) and polypyrrole composite film. Analytical Chemistry. 2009;**81**:4089-4098. DOI: 10.1021/ac900368m

[60] Wei Y, Kong L-T, Yang R, Wang L, Liu J-H, Huang X-J. Electrochemical impedance determination of polychlorinated biphenyl using a pyrenecyclodextrin-decorated single-walled carbon nanotube hybrid. Chemical Communications. 2011;**47**: 5340. DOI: 10.1039/c1cc11267h

[61] Wei Y, Kong L-T, Yang R, Wang L, Liu J-H, Huang X-J. Single-walled carbon nanotube/pyrenecyclodextrin nanohybrids for ultrahighly sensitive and selective detection of *p* -nitrophenol. Langmuir. 2011;**27**:10295-10301. DOI: 10.1021/la201691c

[62] Niu X, Mo Z, Yang X, Shuai C, Liu N, Guo R. Graphene-ferrocene functionalized cyclodextrin composite with high electrochemical recognition capability for phenylalanine enantiomers. Bioelectrochemistry. 2019; **128**:74-82. DOI: 10.1016/j.bioelechem.2019.03.006

[63] Ji J, Qu L, Wang Z, Li G, Feng W, Yang G. A facile electrochemical chiral sensor for tryptophan enantiomers based on multiwalled carbon nanotube/hydroxypropyl-β-cyclodextrin functionalized carboxymethyl cellulose. Microchemical Journal. 2022;**175**:107133. DOI: 10.1016/j.microc.2021.107133

[64] Alam AU, Deen MJ. Bisphenol a electrochemical sensor using graphene oxide and β-cyclodextrin-functionalized multi-walled carbon nanotubes. Analytical Chemistry. 2020;**92**: 5532-5539. DOI: 10.1021/acs.analchem.0c00402

[65] AlRabiah H, Homoda A, Bakheit A, Mostafa GAE. Cyclodextrin potentiometric sensors based on selective recognition sites for procainamide: Comparative and theoretical study. Open Chemistry. 2019;

17:1222-1234. DOI: 10.1515/chem-2019-0131

[66] Mahmoud Mostafa S. Novel potentiometric sensors based on *β*-cyclodextrin and dibenzo 18-crown-6 Ionophores/mesoporous silica nanoparticles for clidinium determination. International Journal of Electrochemical Science. 2020: 3347-3364. DOI: 10.20964/2020.04.48

[67] Lenik J, Łyszczek R. Functionalized *β*-cyclodextrin based potentiometric sensor for naproxen determination. Materials Science and Engineering: C. 2016;**61**:149-157. DOI: 10.1016/j.msec.2015.12.011

[68] Lenik J, Łyszczek R. A potentiometric sensor for ketoprofen based on a *β*-cyclodextrin derivative. Journal of Analytical Chemistry. 2022;**77**: 246-256. DOI: 10.1134/S1061934822020058

[69] Amorim CG, Araújo AN, Montenegro MCBSM, Silva VL. Cyclodextrin-based potentiometric sensors for midazolam and diazepam. Journal of Pharmaceutical and Biomedical Analysis. 2008;**48**: 1064-1069. DOI: 10.1016/j.jpba.2008.08.012

[70] Khaled E, Hassan HNA, Mohamed GG, Ragab FA, Seleim AEA. *β*-Cyclodextrin-based potentiometric sensors for flow-injection determination of acetylcholines. International Journal of Electrochemical Science. 2010;**5**: 448-458

[71] Mousazadeh H, Pilehvar-Soltanahmadi Y, Dadashpour M, Zarghami N. Cyclodextrin based natural nanostructured carbohydrate polymers as effective non-viral siRNA delivery systems for cancer gene therapy. Journal of Controlled Release. 2021;**330**: 1046-1070. DOI: 10.1016/j.jconrel.2020.11.011

[72] Ogoshi T, Takashima Y, Yamaguchi H, Harada A. Cyclodextrin-grafted poly(phenylene ethynylene) with chemically-responsive properties. Chemical Communications. 2006:3702. DOI: 10.1039/b605804c

[73] Rather IA, Ali R. An efficient and versatile deep eutectic solvent-mediated green method for the synthesis of functionalized coumarins. ACS Omega. 2022;**7**:10649-10659. DOI: 10.1021/acsomega.2c00293

[74] Hossain MA, Mihara H, Ueno A. Novel peptides bearing pyrene and coumarin units with or without *β*-cyclodextrin in their side chains exhibit intramolecular fluorescence resonance energy transfer. Journal of the American Chemical Society. 2003;**125**:11178-11179. DOI: 10.1021/ja036427y

[75] Hossain MA, Mihara H, Ueno A. Fluorescence resonance energy transfer in a novel cyclodextrin–peptide conjugate for detecting steroid molecules. Bioorganic & Medicinal Chemistry Letters. 2003;**13**:4305-4308. DOI: 10.1016/j.bmcl.2003.09.051

[76] Fujimoto K, Muto Y, Inouye M, Duplex-Based ADNA. Tailor-made fluorescent sensor for porphyrin derivatives. Bioconjugate Chemistry. 2008;**19**:1132-1134. DOI: 10.1021/bc800133e

[77] Nazerdeylami S, Ghasemi JB, Mohammadi Ziarani G, Amiri A, Badiei A. Direct monitoring of diclofenac using a supramolecular fluorescent approach based on *β*-cyclodextrin nanosponge. Journal of Molecular Liquids. 2021;**336**:116104. DOI: 10.1016/j.molliq.2021.116104

[78] Gao L, He C. Application of nanomaterials decorated with cyclodextrins as sensing elements for environment analysis. Environmental Science and Pollution Research. 2021;**28**: 59499-59518. DOI: 10.1007/s11356-021-16201-6

[79] Alvi S, Ali R. Synthetic approaches to bowl-shaped π-conjugated sumanene and its congeners. Beilstein Journal of Organic Chemistry. 2020;**16**:2212-2259. DOI: 10.3762/bjoc.16.186

[80] Ali R, Alvi S. The story of π-conjugated isotruxene and its congeners: From syntheses to applications. Tetrahedron. 2020;**76**:131345. DOI: 10.1016/j.tet.2020.131345

[81] Wagay SA, Rather IA, Ali R. Functionalized truxene scaffold: A promising advanced organic material for digital era. ChemistrySelect. 2019;**4**: 12272-12288. DOI: 10.1002/slct.2019 03076

[82] Yuan D-Q, Koga K, Kourogi Y, Fujita K. Synthesis of fullerene–cyclodextrin conjugates. Tetrahedron Letters. 2001;**42**:6727-6729. DOI: 10.1016/S0040-4039(01)01375-2

[83] Ogoshi T, Takashima Y, Yamaguchi H, Harada A. Journal of the American Chemical Society. 2007;**129**: 4878-4879. DOI: 10.1021/ja070457+

[84] Zhao Y-L, Hu L, Stoddart JF, Grüner G. Pyrenecyclodextrin-decorated single-walled carbon nanotube field-effect transistors as chemical sensors. Advanced Materials. 2008;**20**:1910-1915. DOI: 10.1002/adma.200702804

[85] Liu J, Mendoza S, Román E, Lynn MJ, Xu R, Kaifer AE. Cyclodextrin-modified gold nanospheres. Host−guest interactions at work to control colloidal properties. Journal of the American Chemical Society. 1999;**121**:4304-4305. DOI: 10.1021/ja990330n

[86] Liu J, Ong W, Román E, Lynn MJ, Kaifer AE. Cyclodextrin-modified gold nanospheres. Langmuir. 2000;**16**: 3000-3002. DOI: 10.1021/la991519f

[87] Liu J, Alvarez J, Ong W, Kaifer AE. Network aggregates formed by C_{60} and gold nanoparticles capped with γ-cyclodextrin hosts. Nano Letters. 2001;**1**: 57-60. DOI: 10.1021/nl0001813

[88] Tang B, Cao L, Xu K, Zhuo L, Ge J, Li Q, et al. A new nanobiosensor for glucose with high sensitivity and selectivity in serum based on fluorescence resonance energy transfer (FRET) between CdTe quantum dots and Au nanoparticles. Chemistry–A European Journal. 2008;**14**:3637-3644. DOI: 10.1002/chem.200701871

[89] Bindhu MR, Saranya P, Sheeba M, Vijilvani C, Rejiniemon TS, Al-Mohaimeed AM, et al. Functionalization of gold nanoparticles by β-cyclodextrin as a probe for the detection of heavy metals in water and photocatalytic degradation of textile dye. Environmental Research. 2021;**201**: 111628. DOI: 10.1016/j.envres.2021. 111628

[90] Elgamouz A, Nassab C, Bihi A, Mohamad SAI, Almusafri AHSA, Alharthi SS, et al. Encapsulation capacity of β-cyclodextrin stabilized silver nanoparticles towards creatinine enhances the colorimetric sensing of hydrogen peroxide in urine. Nanomaterials. 2021;**11**:1897. DOI: 10.3390/nano11081897

Chapter 4

Cyclodextrins as Supramolecular Hosts for Dye Molecules

Olga Fedorova and Yuri Fedorov

Abstract

Cyclodextrins possess a hydrophobic cavity due to which they are suitable for inclusion of various organic dyes. The complex formation between CDs and dyes has been employed to affect the photophysical and photochemical characteristics of organic dyes such as fluorescence enhancement, charge and proton transfer, energy transfer, photochromic transformations and intramolecular excimer/exciplex formation. This fundamental approach has also potential nanotechnological application in creation of optoelectronic devices. Thus, the fluorescent hybrid nanomaterials consisting of supramolecular assemblies of cyclodextrins with fluorescent dyes can be considered as multivalent scaffolds for the construction of various devices applicable in science and technology, in fluorescence spectroscopic and microscopic techniques providing high sensitivity and imaging of cells with high resolutions. In some cases, fluorescent hybrid materials composed of CD-dye complexes have been successfully used instead of the fluorescent organic molecules in sensing and bioimaging studies.

Keywords: organic dyes, cyclodextrin, complex formation, fluorescence, photophysical properties, photochromism

1. Introduction

Cyclodextrins (CDs) present oligosaccharides (between 973 and 2163 Da) composed from D-glucopyranose units (**Figure 1**) [1]. CDs are commonly classified as α-, β- and γ- CDs containing six, seven and eight glucopyranose units, respectively [2–4]. Cyclodextrins may include molecules whose size and polarity are compatible with their lipophilic interior cavity. The interaction forces participated in complexation are dipole-dipole interaction, electrostatic interaction, van der Waals interaction, dispersion forces, hydrophobic interaction, and conformational strain reduction [5, 6].

Complexation reactions involving cyclodextrins are important for two purposes: research investigations of the molecule interactions with cyclodextrin, and applied technologies (pharmaceutical chemistry, food, cosmetic, chemical synthesis and catalysis) [7–9]. It is important to note that the ability of cyclodextrins to bind guest molecules in their cavities has been used to affect the photochemical and photophysical properties of organic dyes, such as enhancement of fluorescence and

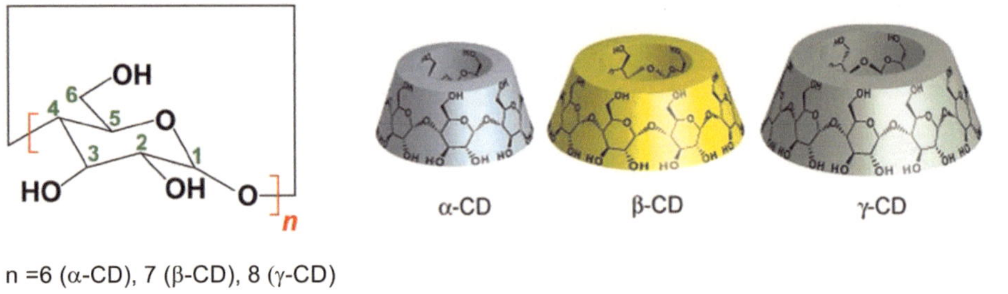

Figure 1.
Structures of α-, β- and γ- CDs.

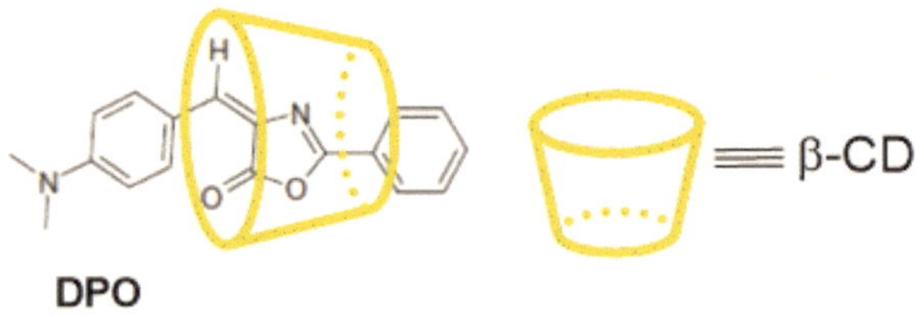

Figure 2.
Structure of dye ***DPO****-β-CD complex.*

phosphorescence, intramolecular charge transfer, intermolecular hydrogen bonding, intramolecular excited proton transfer, intermolecular excimer/exciplex formation. [10–16].

2. Effect of CD-dye complex formation on dye photophysical properties

For dye possessing high extinction coefficient and large difference in dipole moment between ground and singlet excited state, the photo-induced charge transfer (ICT) can be proposed. For such dyes, fluorescence quantum yield and singlet excited state lifetime are sensitive to the polarity of the solvent. Also dye–solvent interaction such as hydrogen bonding can be realized for polar dyes [17]. 4-(p-N,N-Dimethyl-aminophenylmethylene-2-phenyl-5-oxazolone (**DPO**) is the polar dye demonstrated ICT upon photoirradiation. The blue shift found in the emission spectrum of dye **DPO** on adding β-CD points that the molecule is buried in the hydrophobic cavity of β-CD (**Figure 2**). In the resulting dye complex, restrictions on molecular movement appear, which causes the processes of nonradiative deactivation. Thus, restrictions on the motion of the dye molecule in the complex with β-CD lead to a noticeable increase in the fluorescence quantum yield. In this work, it was shown that the dye **DPO** in ICT excited state becomes more polar which results in destabilization of CD-**DPO** complex.

Similar effect on fluorescence has been shown upon complex formation of 2-styrylbenzothiazole containing 15-crown-5 ether fragment (**CSB**) with hydroxypropyl-β-CD (HP-β-CD) (**Figure 3**) [18]. The fluorescence quantum yield of **CSB** is enhanced by 5 times in HP-β-CD relative to water. **CSB** in the HP-β-CD cavity takes place very probably along the molecular axis with a nearly anti-parallel dipole-dipole orientation.

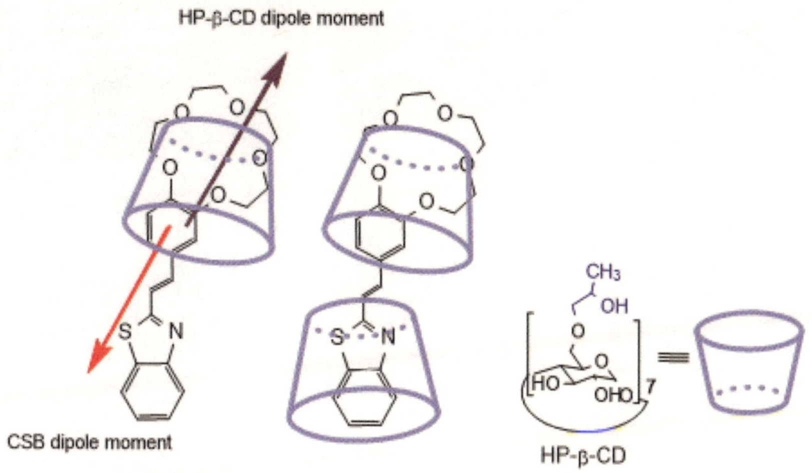

Figure 3.
*Structure of dye **CSB**-β-CD complex.*

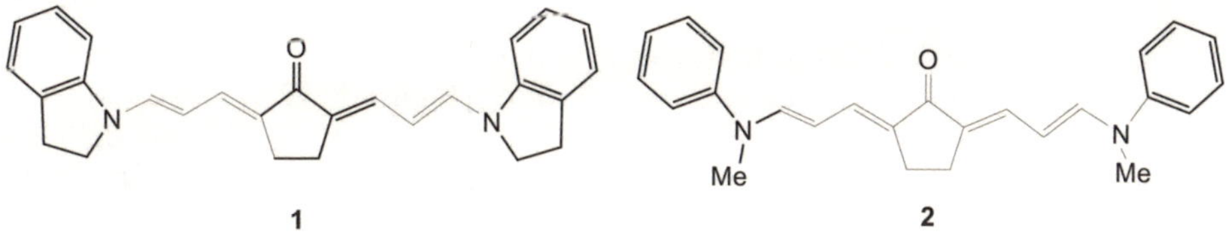

Figure 4.
Structures of dyes 1, 2.

Optical characteristics of ketocyanine dye molecules **1**, **2** (see **Figure 4**) are dependent on the immediate environment [19]. Thus, in an alcohol solvent, the anisotropy of dye fluorescence is inferred. The dyes are only slightly soluble in water. Increased solubility in β-CD indicates dye–β-CD interaction. As shown by investigation of β-CD - ketocyanine dyes **1**, **2** complexes, the strong dye–β-CD interaction is accompanied by high values of fluorescence anisotropy. The dyes **1**, **2** interact through the carbonyl part with hydroxyl groups in β-CD, such interactions are important for stabilization of complex.

The complex formation of 4-amino-2,5-dimethoxybenzanilide (Blue RR (**FBRR**)) and 4-amino-5-methoxy-2-methylbenzanilide (violet B, **FVB**) with hydroxypropyl-α-cyclodextrin (HP-α-CD) and HP-β-CD was studied in [20]. **FVB** and **FBRR** in HP-α-CD demonstrated lack of complexation probably due to smaller cavity size. The deep penetration of the **FVB/FBRR** in HP-β-CD than that of HP-α-CD may be due to the difference in size, also the strong hydrogen bonding of the alcoholic OH on the CD ring with the CONH group of the guests was suggested. The dual fluorescence of both dyes was observed through the normal emission around 350 nm and the very low TICT band around 455 nm. Upon addition of HP-α-CD, the emission slightly increased, and the intensity ratio of the TICT band and the normal band *Ia/Ib* was the same. When the concentration of HP-α-CD increased, the *Ia/Ib* ratio decreased. It has been shown that FVB/FBRR TICT radiation significantly affects the formation of inclusion complexes with different geometries. The explanation was done in terms of differences in the internal diameters of both CD cavities, as well as the change in charge distribution in **FVB** and **FBRR** in CD complexes.

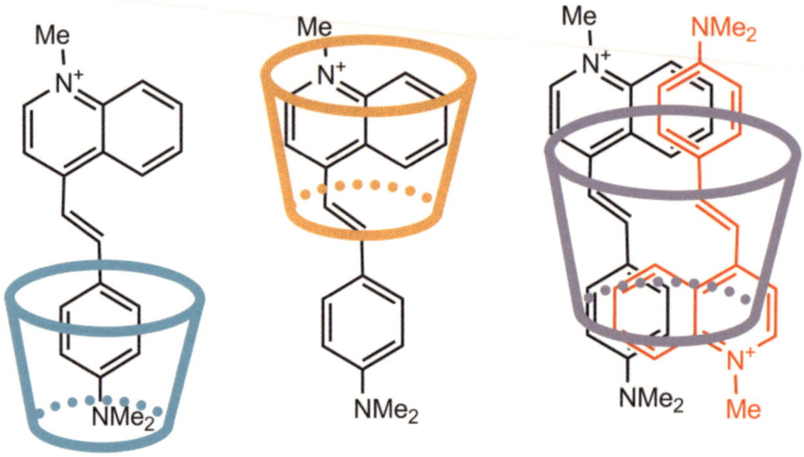

3·α-CD 3·β-CD 3·γ-CD

Figure 5.
Structures of complex between dye 3 and α-, β-, γ-CDs.

1-Methyl-4-(4-aminostyryl) quinolinium iodide **3** forms inclusion complexes with α,β,γ-cyclodextrins in the ground and excited states (**Figure 5**) [21]. The fluorescence quantum yields (Q_{fl}) were 0.043 in water, 0.06 in γ-CD, 0.08 in α-CD, and 0.38 in β-CD. The increase in the fluorescence Q_{fl} indicates better accommodation of the dye in β-CD compared to the other cavities. The fluorescence spectra showed an additional band at longer wavelength in case of γ-CD. This may be attributed to an excimer of two adjacent **3** molecules.

The interesting observations have come out when **Coumarin** dye forms the complex with electron transfer (ET) with N,N-dimethylanyline (**DMA**) in DMF [22]. The ET occurs in a contact ion pair between **Coumarin** and **DMA**. It was also found that the ET rate decreases in a polar solvent medium. This happens because the hydrogen bonds between the **Coumarin** and **DMA** are partially broken due to the presence of solvent molecules between the reactants. **Coumarin** dye in DMF binds to cyclodextrin molecule to form 1:1 and 2:1 complexes through hydrogen bonding. ET process between **Coumarin**-CD complex and **DMA** was not observed (**Figure 6**).

Dual fluorescence (from TICT and plane molecule) of 4-dimethylaminobenzonitrile (**DMABN**) has been studied in α-cyclodextrin (α-CD) complex [23]. **DMABN** molecules are located in two different positions inside the α-CD cavity

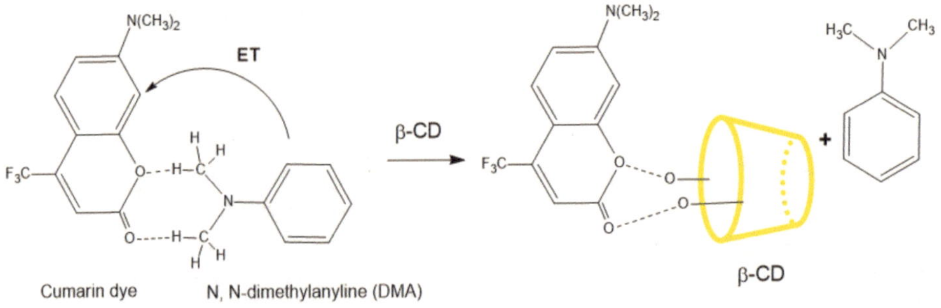

Figure 6.
Structures of complexes between Cumarin dye, DMA and β-CD.

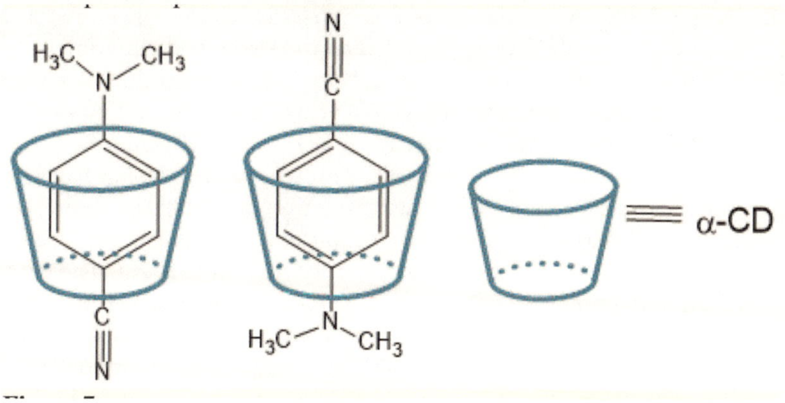

Figure 7.
Structures of complexes between ***DMABN*** *and α-CD.*

(**Figure 7**). The first position is when the dimethylamino group of the **DMABN** molecule is headed towards the larger rim of the α-CD cavity. In this position, amino group is in a slightly polar and slightly rigid environment. In the second position, the dimethylamino group of the **DMABN** molecule is headed towards the smaller rim of the α-CD cavity. In the second position, amino group is in the least polar and most rigid environment. The intensities of both plane and TICT fluorescence bands are enhanced in both types of complexes with α-CD. However, the fluorescent band of plane molecule is more enhanced upon complex formation with α-CD than those of TICT band.

It has been known that the molecule methyl *o*-hydroxy-p-dimethylaminobenzoate (**MHDMAB**) demonstrates triple fluorescence i.e., the normal-locally excited state emission, IF(LE), the intramolecular proton-transfer tautomer emission, IF(IPT), and twisted intramolecular charge-transfer emission, IF(TICT) [24]. It was found that α- and β-cyclodextrins affect both emission modes LE and TICT of the fluorescence spectrum of **MHDMAB** in aqueous solution (**Figure 8**) [24]. This study showed that **MHDMAB** in α-CD and β-CD formed both 1:1 and 1:2 inclusion complexes. The photophysical behavior of **MHDMAB** is modified significantly upon encapsulation of the dye inside β-CD cavities. The short-wavelength emission band of **MHDMAB** in

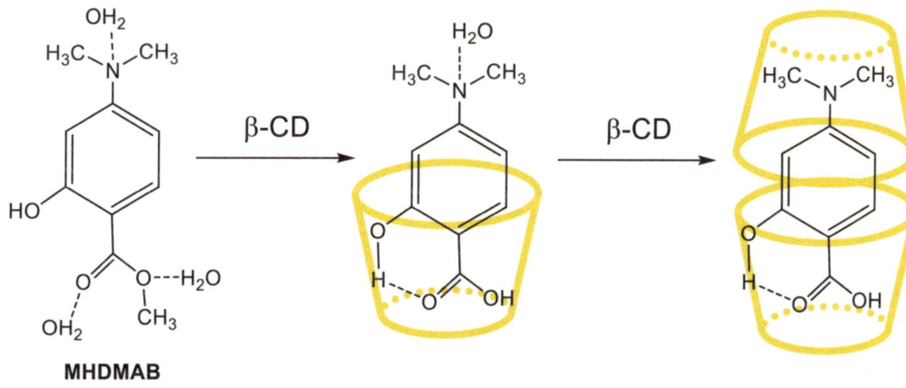

Figure 8.
Structures of complexes between ***MHDMAB*** *and β-CD.*

water is increased as the concentration of β-CD increases, also new emission bands at 450 nm (IPT) and 525 nm (TICT) appeared. The time-resolved experiments gave the fluorescence decay time of the fast component originating from the emission of the hydrogen-bonded complex (**MHDMAB**-H_2O), whereas the decay times of the slow component are related to the fluorescence from IPT and TICT states.

The irradiation of bisstyryl dye **4** with 335 nm light causes the light absorption by the neutral 4-styrylpyridine fragment. The irradiation results in the fluorescence of the positively charged part of the bisstyryl dye in the region of 550 nm (see **Figure 9**). Thus, RET from the neutral to the charged part occurs in dye **1** [25]. The experiments have shown that, in the presence of CB and formation of complex **3**@HP-β-CD, it does not affect the efficiency of the resonant energy transfer process, while binding to CB [7] molecules or simultaneously to HP-β-CD and CB[7] molecules reduce the efficiency of FRET. The observed effect can be explained by the fact that the optical characteristics of styryl fragments also noticeably change during CB[7] encapsulation, and the mutual arrangement of styryl fragments in supramolecular complexes also changes.

Supramolecular systems containing porphyrinoid compounds are of great interest due to such characteristics as high molar absorption in the ultraviolet and visible regions of light, easily changing properties, high chemical stability, and long lifetime of the first excited singlet state [26]. In some porphyrin and phthalocyanine macrocyclic systems, CDs provide the desired supramolecular architecture [27]. Thus, in the porphyrin-CD complexes obtained by Kuroda et al. [28] and Lang et al. [29], electron transfer was discovered. A similar process was found in complexes of adamantaneamine-modified porphyrins with mono-6-*p*-nitrobenzoyl-β-cyclodextrin [30]. Detailed steady-state and time-resolved fluorescence measurements revealed two pathways for electron transfer: dynamic quenching occurring between free donors and free acceptors in solution, and static quenching between donors and acceptors bound in a supramolecular complex.

Other research groups investigated supramolecular assemblies which were composed not only of porphyrins but also of other porphyrinoids, for example, phthalocyanines [31]. One of the most interesting examples was presented as self-assembled complexes containing tetra(*p*-sulfophenyl)porphyrin (**TPPS**) and permethylated-β-CD, conjugated axially to phthalocyanine or subphthalocyanine molecules [31]. The efficient energy transfer from the photoexcited phthalocyanine to a free-base porphyrin occurred comparable to the values found for multiporphyrin arrays linked with covalent bonds.

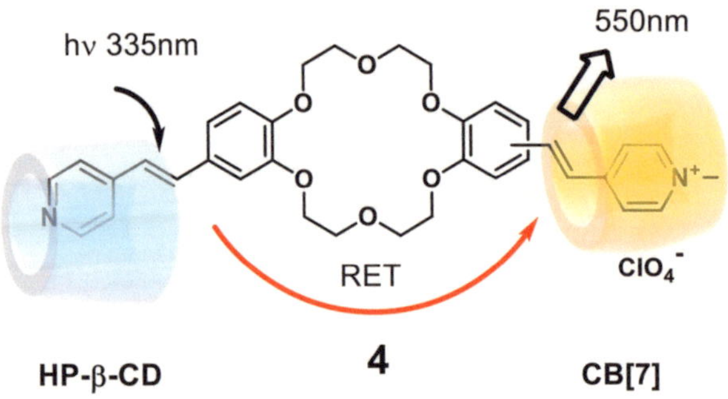

Figure 9.
Complex CB[7]@4@HP-β-CD.

3. Effect of CD-dye complex formation on the dye photochromism

Photochromism is the reversible photoinduced transformation of molecules. Photochromic molecules have been extensively applied as components of nonlinear devices, optical memories, and optical switches [32–34]. To extend the range of commercial applications, photochromes are typically introduced in different materials, including cyclodextrins (CDs) [35–37].

The aim of the work [38] was to study the complex formation of crown-containing styrylheterocycles **5**, **6** with modified cyclodextrin HP-β-CD in aqueous solutions. The investigation of the photochemical reactions of these compounds proceeding in the cavity of cyclodextrin has been carried out (**Figure 10**). It was shown that the process of complex formation causes a significant increase in solubility and results in an intensive luminescent response of styrylheterocycle molecules. In such complexes, the reversible *E–Z* photoisomerization is taking place in aqueous media. The photoisomerization does not cause the destruction of 1:1 complexes, staying guest molecules encapsulated. In opposite, encapsulated 1:2 complex was not found in *Z*-form.

The same styryl dyes **5**, **6** have been exploited as photoactive guests in three-component systems containing both HP-β-CD and cucurbit[7]uril (CB[7]) host molecules [39]. The formation of complexes *E*-**5** and *E*-**6** with HP-β-CD occurs with constants $\log K_{11}$ = 3.58 and $\log K_{11}$ = 3.04 correspondently (**Figure 11**). Starting from the *E*-**5**, *E*-**6**, the phototransformation under light (≥ 320 nm) is observed and includes two consecutive photochemical reactions, an *E-Z* isomerization reaction and a 1-aza-1,3,5-hexatrienic electrocyclic reaction in which the formation of C-N bond was observed. The cyclic product gives stable heteroaromatic cations as a result of elimination of the hydride with atmospheric oxygen (products **7**, **8** (**Figure 11**). The physicochemical analysis of the phototransformation showed that the formation of Z-isomer can occur in HP-β-CD, whereas the formation of heteroaromatic cations **7**, **8** leads to the destruction of HP-β-CD complex (**Figure 11**). The presence of CB[7] in the **5** or **6** solution causes the formation of novel complexes **5**·CB[7] or **6**·CB[7]. Thus, a three-component system involving styryl dye and both HP-β-CD and CB[7] hosts can be switched on by the synchronous host–guest complexation of dye with HP-β-CD or CB[7] by phototransformation of the dye component.

The α-cyclodextrin [2]-rotaxanes have been obtained with alkane-, stilbene- (**Figure 12**) and azobenzene-based axles with different substituents [40–43]. The rotaxanes based on azobenzene and stilbene derivatives were found to demonstrate

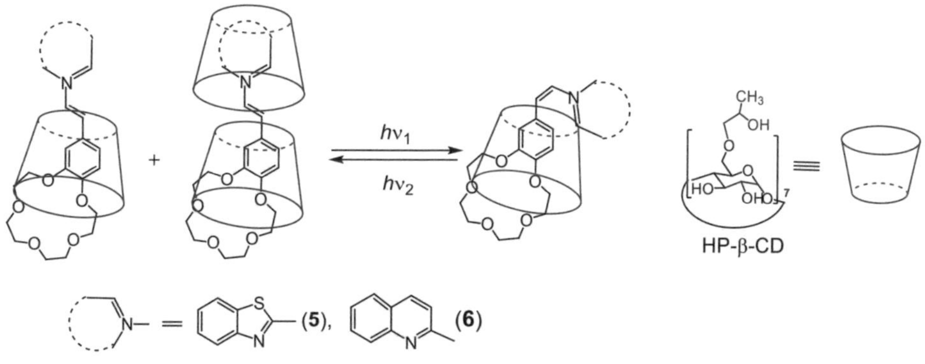

Figure 10.
Photoisomerization of complexes 5, 6 with HP-β-CD.

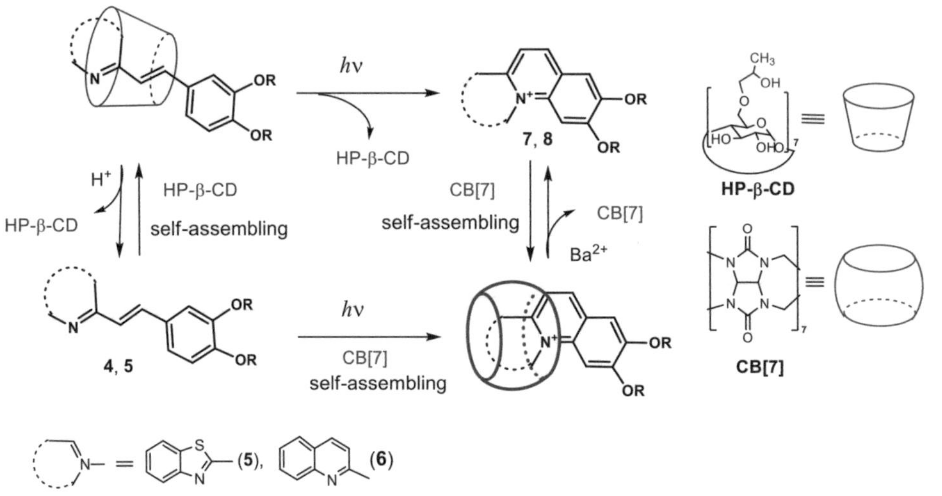

Figure 11.
*Photochemical transformation in three-component systems containing both HP-β-CD, CB[7] and photoresponsive dye 5 or **6**.*

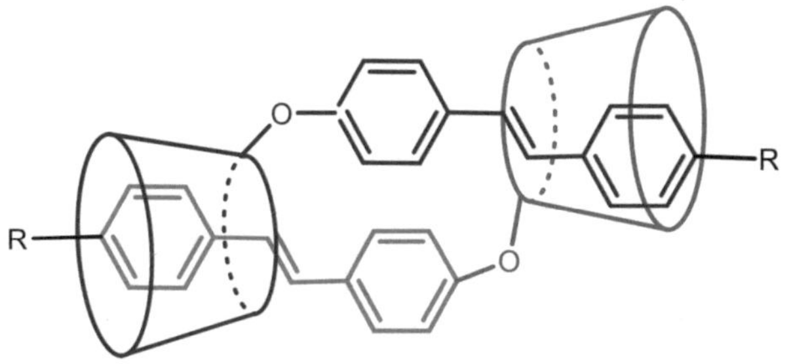

Figure 12.
Structure of α-cyclodextrin [2]-rotaxanes based on stilbene derivatives.

photochemically induced reversible mutual conversion between its *trans*- and *cis*-isomers, resulting in the moving of the cyclodextrin back and forth along the axle. The α-cyclodextrin [2]-rotaxanes behave as a molecular shuttle. This type of light-powered rotaxane exhibits favorable repeatability and presents a novel light-driven molecular machine.

Chen et al. [44] obtained that phthalocyanine containing azabenzene moiety in *trans* form can easily enter into α-CD, but in *cis* form azobenzene moiety is not planar what prevents host-guest interaction between phthalocyanine and α-CD (**Figure 13**).

Mulder et al. studied dithienylethene-tethered β-CD dimers in which the irradiation with UV light caused photochemical ring-closure reaction [45]. Tetra(*p*-sulfophenyl)porphyrin (**TPPS**) was used as a guest for the interaction with dithienylethene-tethered β-CD. It was found that the alternation of UV and visible light irradiations caused a reversible release and uptake of porphyrin.

Synthesis of cyclodextrin polymers using cross-linking agents has been described in literature [46]. Such substituted by CD polymers are called cyclodextrin-based nanosponges [46]. Using this approach, a series of photochromic polymers were prepared by forming various spiropyran (**SP**) inclusion complexes in the CD cavities of the β-CD polymer (**CDP**) (**Figure 14**) [47]. The β-CD is not able to include the **SP**,

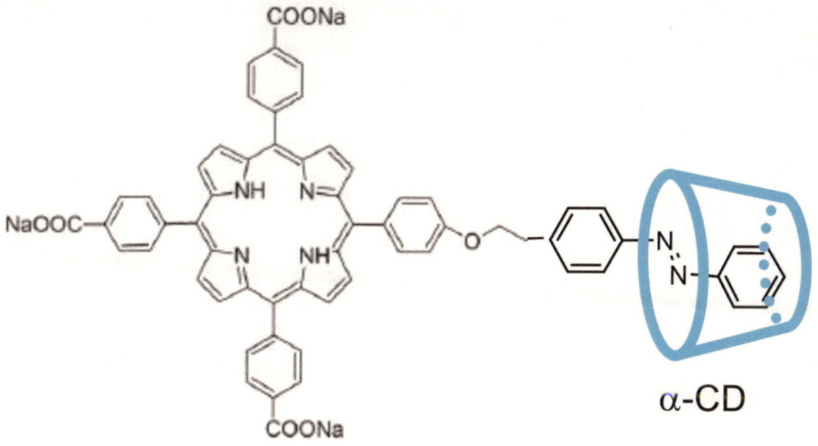

Figure 13.
Structure of α-cyclodextrin complex with phthalocyanine containing azabenzene moiety.

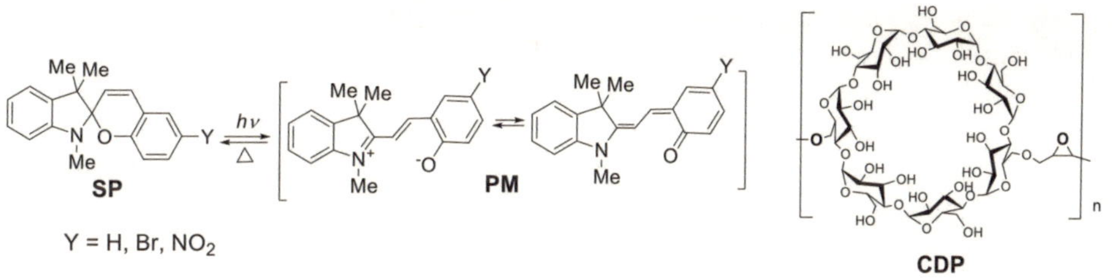

Figure 14.
Photochromic transformation of ***SP*** *to* ***PM*** *and structure of β-CD polymer (****CDP****).*

whereas the photomerocyanine form (**PM**) can bind with β-CD. Indeed, the decolouration rate of **PM** forms is decreased in the presence of β-CD. Consistent with experimental UV/VIS spectra, the quantum chemical calculations provided valuable insight into the substituent and CDP effects on the **PM** decolouration process. It was found that k_{CDPM}/k_{PM} ratio is larger than 20, k is constant of decolouration. Thus, NO_2-**SP** and Br-**SP** exhibit slower decolouration rates in β-CD than alone **SP** because the narrow β-CD cavity hinders deep inclusion of the bulky naphthopyryl moiety.

4. CD-dye complexes in biology and medicine

CD systems have been extensively studied in biology and medicine, because CD counterpart acts like a comparable protein structure, providing the proper environment and arrangement of the substrates. In some cases, the processes taking place in these systems mimic those occurring in living organisms.

A novel approach towards controlled ligand–DNA interactions has been developed based on supramolecular complex of dye **9** and HP-β-CD [48]. Dye **9** is not able to coordinate with DNA (**Figure 15**). The irradiation of encapsulated **9** caused the electrocyclic transformation to product **10** which could not be able to be bound with HP-β-CD but easily interacts with DNA. Thus, the process begins with photoinduced in situ generations of a DNA ligand from the encapsulated styryl heterocycle,

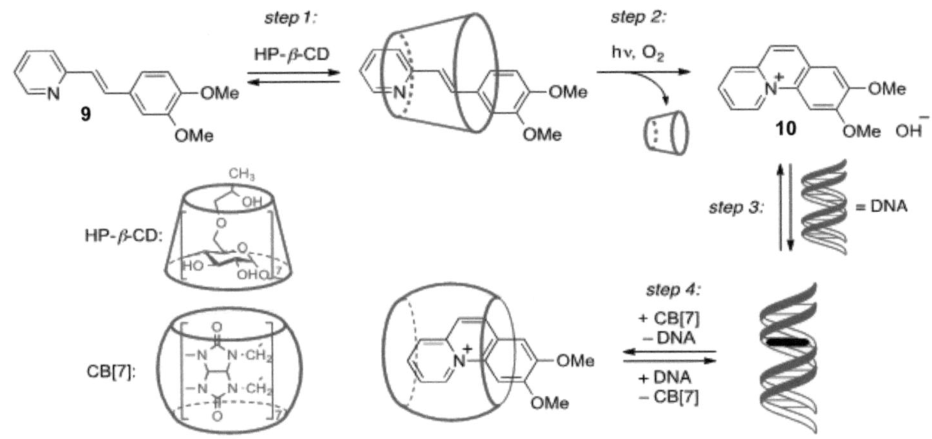

Figure 15.
*Association and redistribution equilibria of ligands **9** and **10** in the presence of hosts.*

continues with the association of the ligand with the nucleic acid, and ends with the removal of the bound ligand from the DNA binding site using CB[7]. Despite the simultaneous presence of several host molecules, HP-β-CD, CB[7] and DNA, each step of the transformation cascade is not affected by the presence of other components. It is important to note that the phototransformation of precursor **9** into DNA intercalator **10** inside the cyclodextrin cavity significantly increases the biocompatibility of the method.

Molecules like cyclodextrins can be applied to solve both the solubility and the toxicity of the fluorescent dyes using in fluorescence imaging techniques to visualize and monitor specific biological targets or processes in living systems. Thus, Alexandru Rotaru and co-authors demonstrated a low level of fluorescent dye **11** toxicity (**Figure 16**) by the formation of cyclodextrin inclusion complex resulting in the successful application in cell staining [49]. Applications of this type of compound are limited due to high toxicity and water solubility problems. The addition of β-CD to dye **11** solutions in ratio of 3:1 results in dye being soluble in water. Also, fluorescent indolizinyl-pyridinium salt/β-cyclodextrin inclusion complexes demonstrated absence of cytotoxicity. Due to found in experiments cellular permeability, long-lived intracellular fluorescence and selective accumulation within acidic organelles, the dye **11** can be identified as remarkable candidate for intracellular labelling of acidic organelles (lysosomes or mitochondria).

Nanosponges prepared based on β-cyclodextrin and diphenyl carbonate have the capacity to interact with small molecules in their matrix [50]. Flavonoid quercetin was loaded into such nanosponges (**Figure 17**) [51]. The dissolution of the quercetin

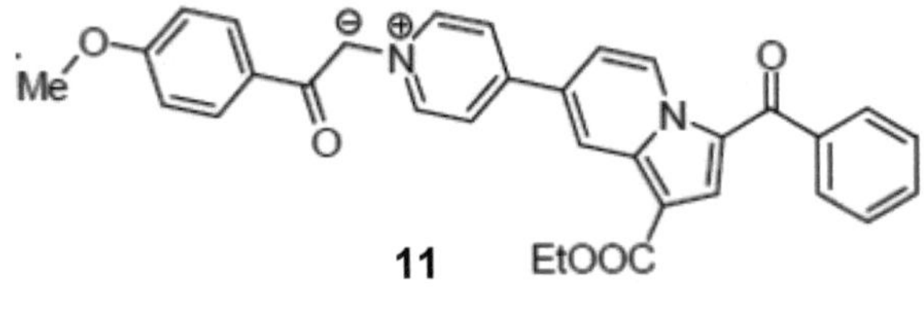

Figure 16.
*Structure of fluorescent dye **11**.*

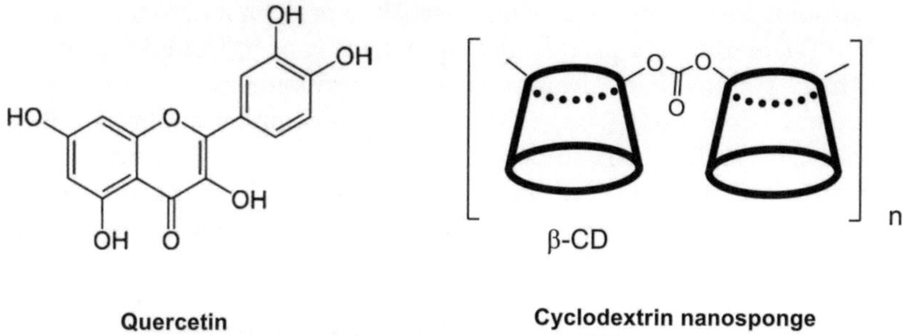

Figure 17.
Structures of quercetin and nanosponge.

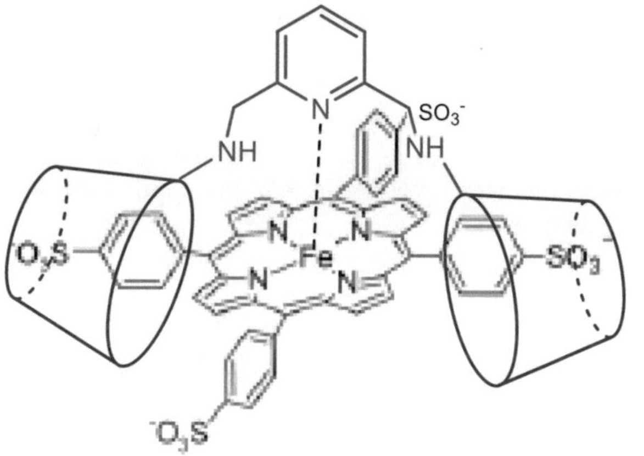

Figure 18.
*Complex of substituted β-CD dimer and **Fe(II)TPPS**.*

nanosponges was significantly higher compared with the pure drug. The stability of encapsulated quercetin nanosponge was markedly improved. In addition, the antioxidant activity of the quercetin in composition of nanosponges was higher than pure quercetin.

Supramolecular ensembles of porphyrinoid-CD are formed both through covalent binding and through the formation of inclusion complexes [27]. Such systems are biomimetics that mimic the natural breakdown of carotenoids [52], cytochrome P450 mediated hydroxylation [53], and oxygen binding by hemoglobin [54].

The system reversibly bound O_2 was proposed by Zhou and Groves [55], it is based on self-assembly of the Fe(II)-tetra(*p*-sulfophenyl)porphyrin (**Fe(II)TPPS**) and β-CD derivative having pyridylmethyl moiety and PEG groups. Pyridyl moiety served as a ligand for **Fe(II)TPPS** and PEG chains spanned over the porphyrin surface, protecting the second binding site. The binding of O_2 and CO was proved by optical method.

Another analytical method has been used by Koji Kano and his coworkers [56]. Their systems contained substituted β-CD dimers and **Fe(II)TPPS** (**Figure 18**). 4-Sulfonatophenyl groups of porphyrins were embedded in β-CD moieties of the dimer, whereas the pyridine linker coordinated the Fe(II) central ion. The hydrophobic environment within the Fe(II) ionic centre of the supramolecular complex was crucial for the efficient O_2 binding, this is why the affinity of such complexes to O_2 is high and stabilities of O_2 adducts are significant.

Porphyrinoids are widely used as photosensitizers in photodynamic therapy (PDT), the PDT method is a promising way to treat cancer. Complexation with CD improves the photosensitizing properties of porphyrinoids since an increase in their quantum yield of singlet oxygen is observed in such complexes. This fact is of great importance for PDT [27]. The complex formation with HP-β-CD improves the efficacy of PDT for the treatment of G361 malign melanoma by using zinc-tetra(*p*-sulfophenyl)porphyrin **ZnTPPS4** [57]. Thus, after 24 h incubation of cell cultures with 10^{-1} M **ZnTPPS4** and 1 mM HP-β-CD, the cells were irradiated for 7.5 min at the total irradiation dose of 12.5 J cm^{-2} which gave rise to DNA damage.

Innovative drug delivery system was proposed based on gold nanoparticles covered by cationic poly(cyclodextrin) (P(CD+)) and alginate (alg−) layers [58]. 4-Hydroxy-tamoxifen was placed in the nanocapsules' shell via inclusion with the cyclodextrin cavities. It was also demonstrated that 4-hydroxy tamoxifen can be efficiently delivered to podocytes *in vitro* using CD-containing nanocapsules as carriers.

Tetrazines functionalized with adamantane groups and naphthalimide antennas can form supramolecular complex with β-cyclodextrin (β-CD) in aqueous solutions [59]. The organic anchoring groups and the tetrazine itself fit well the requirements for cavity cyclodextrin inclusion. This approach was applied for development of biosensors with electrochemical and fluorescence properties [60]. Tetrazine derivatives were immobilized at the electrogenerated polypyrrole-β-CD film through the host-guest interactions between tetrazine derivatives and β-CD. This new original molecular architecture allows the immobilization of glucose oxidase modified by β-CD (**Figure 19**). The absorption band at 425 nm in UV–Vis spectra recorded for ITO

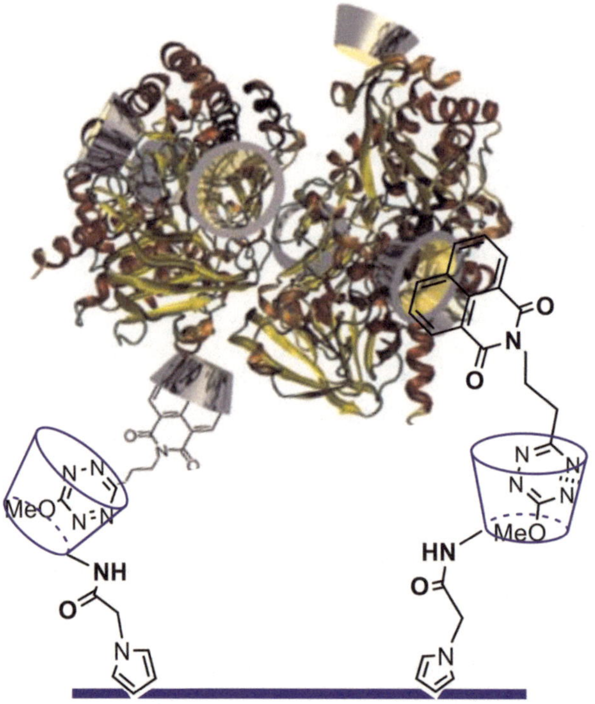

Figure 19.
ITO electrode covered by assembly containing glucose and naphthalimide fragment.

electrodes belonging to naphthalimide fragment confirmed the formation of assembly. The oxidation of glucose in presence of oxygen with the concomitant production of H_2O_2 was investigated by electrochemistry. The prepared electrodes were thus maintained at 0.7 V in presence of glucose to detect, through its oxidation, the enzymatically generated H_2O_2.

5. Conclusion

The examples of complexes of photoactive compounds with CD container molecules presented in this chapter are hybrid materials created by combining two (or more) different elements of a different nature. In this context, they can be considered as a very large and heterogeneous class of materials. Such materials include molecular and supramolecular assembled materials, polymers, or nanosized objects, nanostructured and hybrid architectures with organic or biological characteristics. Such organic hybrid systems combine particular properties of the components, which explains the wide range of properties exhibited by the systems and the great possibilities in the development of methods for their synthesis.

Organic hybrid materials composed of cyclodextrin receptors and photoactive components, provide a great opportunity for improving of photophysical characteristics, increasing functionality, and extending the field of application. Supramolecular functionalization of photoactive CD molecules, which play the role of platforms for the immobilization of bioelements such as enzymes, antibodies, nucleic acids (DNA, RNA, microRNA), and other functional groups, is of great importance for the manufacture of analytical devices for biosensors. This approach can also be used to obtain novel hybrid organic photoactive materials.

Acknowledgements

This work was supported by the Ministry of Science and Higher Education of the Russian Federation (Contract/agreement No. 075-00697-22-00).

Author details

Olga Fedorova* and Yuri Fedorov
A.N. Nesmeyanov Institute of Organoelement Compounds of Russian Academy of Sciences, Moscow, Russia

*Address all correspondence to: fedorova@ineos.ac.ru

References

[1] Sherje AP, Dravyakar BR, Kadam D, Jadhav M. Cyclodextrin-based nanosponges: A critical review. Carbohydrate Polymers. 2017;**173**:37-49. DOI: 10.1016/j.carbpol.2017.05.086

[2] Bibby DC, Davies NM, Tucker IG. Mechanisms by which cyclodextrins modify drug release from polymeric drug delivery systems. International Journal of Pharmaceutics. 2000;**197**: 1-11. DOI: 10.1016/s0378-5173(00) 00335-5

[3] Mura P. Analytical techniques for characterization of cyclodextrin complexes in aqueous solution: A review. Journal of Pharmaceutical and Biomedical Analysis. 2014;**101**:238-250. DOI: 10.1016/j.jpba.2014.02.022

[4] Radu C, Parteni O, Ochiuz L. Application of cyclodextrins in medical textiles - review. Journal of Controlled Release. 2016;**224**:146-157. DOI: 10.1016/j.jconrel.2015.12.046

[5] Liu L, Guo QX. The driving forces in the inclusion complexation of cyclodextrins. Journal of Inclusion Phenomena and Macrocyclic Chemistry. 2002;**42**:1-14. DOI: 10.1023/A: 1014520830813

[6] Li S, Purdy WC. Cyclodextrins and their applications in analytical chemistry. Chemical Reviews. 1992;**92**: 1457-1470. DOI: 10.1021/cr00014a009

[7] Rekharsky MV, Inoue Y. Complexation thermodynamics of Cyclodextrins. Chemical Reviews. 1998; **98**:1875-1918. DOI: 10.1021/cr970015o

[8] Szejtli J. Introduction and general overview of Cyclodextrin chemistry. Chemical Reviews. 1998;**98**:1743-1754. DOI: 10.1021/cr970022c

[9] Jin Z, editor. Cyclodextrin Chemistry: Preparation and Application. Singapore: World Scientific Publishing Co. Pte. Ltd; 2018. p. 292. DOI: 10.1142/10701

[10] Krishnamoorthy G, Dogra SK. Twisted intramolecular charge transfer of 2-(4‘-N,N-Dimethylaminophenyl) pyrido[3,4-d]imidazole in Cyclodextrins: Effect of pH. The Journal of Physical Chemistry. A. 2000;**104**: 2542-2551. DOI: 10.1021/jp992792t

[11] Kim Y, Yoon M, Kim D. Excited-state intramolecular proton transfer coupled-charge transfer of p-N,N-dimethylaminosalicylic acid in aqueous β-cyclodextrin solutions. Journal of Photochemistry and Photobiology A: Chemistry. 2001;**138**:167-175. DOI: 10.1016/S1010-6030(00)00404-4

[12] Dash N, Chipem FAS, Swaminathan R, Krishnamoorthy G. Hydrogen bond induced twisted intramolecular charge transfer in 2-(4′-N,N-dimethylaminophenyl) imidazo[4,5-b]pyridine. Chemical Physics Letters. 2008;**460**:119-124. DOI: 10.1016/j.cplett.2008.05.092

[13] Panja S, Chakravorti S. Photophysics of 4-(N,N-dimethylamino) cinnamaldehyde/α-cyclodextrin inclusion complex. Spectrochimica Acta A. 2002;**58**:113-122. DOI: 10.1016/ S1386-1425(01)00522-4

[14] Banu HS, Pitchumani K, Srinivasan C. Dual emission from 4-dimethylaminobenzonitrile in cyclodextrin derivatives. Journal of Photochemistry and Photobiology A: Chemistry. 2000;**131**:101-110. DOI: 10.1016/S1010-6030(99)00249-X

[15] Chakraborty A, Guchhait N. Inclusion complex of charge transfer probe 4-amino-3-methyl benzoic acid

methyl ester (AMBME) with β-CD in aqueous and non-aqueous medium: Medium dependent stoichiometry of the complex and orientation of probe molecule inside β-CD nanocavity/. Journal of Inclusion Phenomena and Macrocyclic Chemistry. 2008; **62**:91-97. DOI: 10.1007/s10847-008-9442-4

[16] Sen P, Roy D, Mondal SK, Sahu K, Ghosh S, Bhattacharyya K. Fluorescence anisotropy decay and solvation dynamics in a Nanocavity: Coumarin 153 in methyl β-Cyclodextrins. The Journal of Physical Chemistry. A. 2005;**109**:9716-9722. DOI: 10.1021/jp051607a

[17] Asiri AM, El-Daly SA, Khan SA. Spectral characteristics of 4-(p-N,N-dimethyl-aminophenylmethylene)-2-phenyl-5-oxazolone (DPO) in different media. Spectrochimica Acta Part A: Molecular and Biomolecular Spectroscopy. 2012;**95**:679-684. DOI: 10.1016/j.saa.2012.04.077

[18] Fedorov YV, Fedorova OA, Andryukhina EN, Gromov SP, Alfimov MV, Aaron JJ. Guest–host interactions between crown-containing 2-Styrylbenzothiazole and HP-β-CD. Journal of Inclusion Phenomena and Macrocyclic Chemistry. 2004;**49**: 283-289. DOI: 10.1023/B: JIPH.0000048316.04426.f6

[19] Das PK, Pramanik R, Banerjee D, Bagchi S. Studies of solvation of ketocyanine dyes in homogeneous and heterogeneous media by UV:Vis spectroscopic method. Spectrochimica Acta Part A. 2000;**56**:2763-2773. DOI: 10.1016/S1386-1425(00)00321-8

[20] Jenita MJ, Venkatesh G, Subramanian VK, Rajendiran N. Twisted intramolecular charge transfer effects on fast violet B and fast blue RR: Effect of HP-α- and HP-β-cyclodextrins. Journal of Molecular Liquids. 2013;**178**:160-167. DOI: 10.1016/j.molliq.2012.11.033

[21] El-Sayed AM, Al-Sherbini. Spectroscopic study of the inclusion complexes of 1-methyl-4-[4′-aminostyryl] quinolinium iodide with α,β,γ-cyclodextrins in the ground and excited states. Microporous and Mesoporous Materials. 2005;**85**:25-31. DOI: 10.1016/j.micromeso.2005.06.016

[22] Chakraborty A, Seth D, Chakrabarty D, Sarkar N. Photoinduced intermolecular electron transfer from dimethyl aniline to 7-amino Coumarin dyes in the surface of -cyclodextrin. Spectrochimica Acta Part A. 2006;**64**: 801-808. DOI: 10.1016/j.saa.2005.08.007

[23] Al-Hassan KA, Klein UKA, Suwaiyan A. Normal and twisted intramolecular charge-transfer fluorescence of 4-dimethylaminobenzonitrile in α-cyclodextrine cavities. Chemical Physics Letters. 1993;**212**:581-587. DOI: 10.1016/0009-2614(93)85489-B

[24] Lazarowska A, Jozefowicz M, Heldt JR, Heldt J. Spectrochim. Acta A: Molecular and Biomolecular Spectroscopy. 2012;**86**:481-489. DOI: 10.1016/j.saa.2011.10.072

[25] Fedorova OA, Chernikova EY, Tkachenko SV, Grachev AI, Godovikov IA, Fedorov YV. Self-sorting processes in a stimuli-responsive supramolecular systems based on cucurbituril, cyclodextrin and bisstyryl guests. Journal of Inclusion Phenomena and Macrocyclic Chemistry. 2019;**94**: 201-210. DOI: 10.1007/s10847-019-00900-2

[26] Kim JH, Nam DH, Park CB. Nanobiocatalytic assemblies for artificial photosynthesis. Current Opinion in

Biotechnology. 2014;**28**:1-9. DOI: 10.1016/j.copbio.2013.10.008

[27] Kryjewskia M, Goslinskib T, Mielcareka J. Functionality stored in the structures of cyclodextrin – Porphyrinoid systems. Coordination Chemistry Reviews. 2015;**300**: 101-120. DOI: 10.1016/j.ccr.2015. 04.009

[28] Kuroda Y, Ito M, Sera T, Ogoshi H. Controlled electron transfer between cyclodextrin-sandwiched porphyrin and quinones. Journal of the American Chemical Society. 1993;**115**: 7003-7004. DOI: 10.1021/ja000 68a080

[29] Lang K, Král V, Kapusta P, Kubát P, Vašek P. Photoinduced electron transfer within porphyrin–cyclodextrin conjugates. Tetrahedron Letters. 2002; **43**:4919-4922. DOI: 10.1016/ S0040-4039(02)00954-1

[30] Wang YH, Zhu MZ, Ding XY, Ye JP, Liu L, Guo QX. Photoinduced electron transfer between Mono-6-p-nitrobenzoyl-β-cyclodextrin and Adamantanamine-Cn-porphyrins. The Journal of Physical Chemistry. B. 2003; **107**:14087-14093. DOI: 10.1021/ jp0347419

[31] Kano K, Nishiyabu R, Yamazaki T, Yamazaki I. Convenient scaffold for forming Heteroporphyrin arrays in aqueous media. Journal of the American Chemical Society. 2003;**125**: 10625-10634. DOI: 10.1021/ja035055q

[32] Chen S, Guo Z, Zhu S, Shi WE, Zhu W. A multiaddressable photochromic bisthienylethene with sequence-dependent responses: Construction of an INHIBIT logic gate and a keypad lock. ACS Applied Materials & Interfaces. 2013;**5**: 5623-5629. DOI: 10.1021/am4009506

[33] Gentili PL, Dolnik M, Epstein IR. 'Photochemical oscillator': Colored hydrodynamic oscillations and waves in a photochromic system. Journal of Physical Chemistry C. 2014;**118**:598-608. DOI: 10.1021/jp407393h

[34] Gao R, Cao D, Guan Y, Yan D. Flexible self-supporting nanofibers thin films showing reversible photochromic fluorescence. ACS Applied Materials & Interfaces. 2015;**7**:9904-9910. DOI: 10.1021/acsami.5b01996

[35] Elsässer C, Vüllings A, Karcher M, Fumagalli P. Photochromism of spiropyran-cyclodextrin inclusion complexes on Au(111). Journal of Physical Chemistry C. 2009;**113**: 19193-19198. DOI: 10.1021/jp810474v

[36] Burke K, Riccardi C, Buthelezi T. Thermosolvatochromism of nitrospiropyran and merocyanine free and bound to cyclodextrin. The Journal of Physical Chemistry. B. 2012;**116**: 2483-2491. DOI: 10.1021/jp208023r

[37] Zhang SX, Fan MG, Liu YY, Ma Y, Zhang GJ, Yao JN. Inclusion complex of spironaphthoxazine with c-cyclodextrin and its photochromism study. Langmuir. 2007;**23**:9443-9446. DOI: 10.1021/ la700252u

[38] Tkachenko SV, Chernikova EY, Gulakova EN, Godovikov IA, Fedorov YV, Fedorova OA. Photoisomerization of crown containing Styrylbenzothiazole and Styrylquinoline in complexes with Hydroxypropyl-β-cyclodextrin. Protection of Metals and Physical Chemistry of Surfaces. 2013;**49**: 181-188. DOI: 10.1134/ S2070205113020068

[39] Fedorov YV, Tkachenko SV, Chernikova EY, Godovikov IA, Fedorova OA, Isaacs L. Photoinduced guest transformation promotes

translocation of guest from hydroxypropyl-β-cyclodextrin to cucurbit[7]uril. Chemical Communications. 2015;**51**:1349-1352. DOI: 10.1039/C4CC08474H

[40] Dawson RE, Lincoln SF, Easton CJ. The foundation of a light driven molecular muscle based on stilbene and α-cyclodextrin. Chemical Communications. 2008;**34**:3980-3982. DOI: 10.1039/B809014A

[41] Dawson RE, Maniam S, Lincoln SF, Easton CJ. Synthesis of α-cyclodextrin [2]-rotaxanes using chlorotriazine capping reagents. Organic & Biomolecular Chemistry. 2008;**6**: 1814-1821. DOI: 10.1039/B802229A

[42] Cheetham AG, Claridge TDW, Anderson HL. Metal-driven ligand assembly in the synthesis of cyclodextrin [2] and [3]rotaxanes. Organic & Biomolecular Chemistry. 2007;**5**: 457-462. DOI: 10.1039/B616621K

[43] Chen Z, Dong S, Zhong C, Zhang Z, Niu L, Li Z, et al. Photoswitching of the third-order nonlinear optical properties of azobenzene-containing phthalocyanines based on reversible host–guest interactions. Journal of Photochemistry and Photobiology A: Chemistry. 2009;**206**:213-219. DOI: 10.1016/j.jphotochem.2009.07.005

[44] Mulder A, Juković A, van Leeuwen FWB, Kooijman H, Spek AL, Huskens J, et al. Photocontrolled release and uptake of a porphyrin guest by dithienylethene-tethered beta-cyclodextrin host dimers. Chemistry - A European Journal. 2004;**10**:1114-1123. DOI: 10.1002/chem.200305567

[45] Gao C, Ma X, Zhang Q, Wang Q, Qu D, Tian H. A light-powered stretch–contraction supramolecular system based on cobalt coordinated [1]rotaxane. Organic & Biomolecular Chemistry. 2011;**9**:1126-1132. DOI: 10.1039/C0OB00764A

[46] Trotta F, Zanetti M, Cavalli R. Cyclodextrin based nanosponges as drug carriers. Beilstein Journal of Organic Chemistry. 2012;**8**:2091-2099. DOI: 10.3762/bjoc.8.235

[47] Wang LF. Application of response surface methodology for exploring β-cyclodextrin effects on the decoloration of spiropyran complexes. Chemical Physics Letters. 2016;**662**:296-305. DOI: 10.1016/j.cplett.2016.09.068

[48] Berdnikova DV, Aliyeu TM, Paululat T, Fedorov YV, Fedorova OA, Ihmels H. DNA–ligand interactions gained and lost: Light-induced ligand redistribution in a supramolecular cascade. Chemical Communications. 2015;**51**:4906-4909. DOI: 10.1039/C5CC01025J

[49] Pricope G, Ursu EL, Sardaru M, Cojocaru C, Clima L, Marangoci N, et al. Novel cyclodextrin-based pH-sensitive supramolecular host-guest assembly for staining acidic cellular organelles. Polymer Chemistry. 2018;**9**:968-975. DOI: 10.1039/C7PY01668A

[50] Trotta F, Caldera F, Dianzani C, Argenziano M, Barrera G, Cavalli R. Glutathione bioresponsive cyclodextrin nanosponges. ChemPlusChem. 2016;**81**: 439-443. DOI: 10.1002/cplu.201500531

[51] Selvamuthukumar AS. Fabrication of cyclodextrin nanosponges for quercetin delivery: Physicochemical characterization, photostability, and antioxidant effects. Journal of Materials Science. 2014;**49**:8140-8153. DOI: 10.1007/s10853-014-8523-6

[52] French RR, Holzer P, Leuenberger M, Nold MC, Woggon WD.

A supramolecular enzyme model catalyzing the central cleavage of carotenoids. Journal of Inorganic Biochemistry. 2002;**88**:295-304. DOI: 10.1016/s0162-0134(01)00363-4

[53] Fang Z, Breslow R. A thiolate ligand on a cytochrome P-450 mimic permits the use of simple environmentally benign oxidants for biomimetic steroid hydroxylation in water. Bioorganic & Medicinal Chemistry Letters. 2005;**15**: 5463-5466. DOI: 10.1016/j.bmcl.2005.08.090

[54] Jiang T, Lawrence DS. Sugar-coated Metalated macrocycles. Journal of the American Chemical Society. 1995;**117**: 1857-1858. DOI: 10.1021/ja00111a035

[55] Komatsu T, Hayakawa S, Tsuchida E, Nishide H. Meso-Tetrakis [o-(N-methyl)pyridinium]porphyrin ensembles with axially coordinated cyclodextrin-penetrating phenethylimidazole: Reversible dioxygen-binding in aqueous DMF solution. Chemical Communications. 2003;**1**:50-51. DOI: 10.1039/B208882G

[56] Kano K, Itoh Y, Kitagishi H, Hayashi T, Hirota S. A supramolecular receptor of diatomic molecules (O_2, CO, NO) in aqueous solution. Journal of the American Chemical Society. 2008;**130**:8006-8015. DOI: /10.1021/ja8009583

[57] Kolarova H, Macecek J, Nevrelova P, Huf M, Tomecka M, Bajgar R, et al. Photodynamic therapy with zinc-tetra(p-sulfophenyl)porphyrin bound to cyclodextrin induces single strand breaks of cellular DNA in G361 melanoma cells. Toxicology in Vitro. 2005;**19**:971-974. DOI: 10.1016/j.tiv.2005.06.015

[58] Belbekhouche S, Oniszczuk J, Pawlak A, Joukhar IE, Goffin A, Varrault G, et al. Cationic poly (cyclodextrin)/alginate nanocapsules: From design to application as efficient delivery vehicle of 4-hydroxy tamoxifen to podocyte in vitro. Colloids Surfaces B Biointerfaces. 2019;**179**:128-135. DOI: 10.1016/j.colsurfb.2019.03.060

[59] Zhong C, Mu T, Wang L, Fu E, Qin J. Unexpected fluorescent behavior of a 4-amino-1,8-naphthalimide derived β-cyclodextrin: Conformation analysis and sensing properties. Chemical Communications. 2009;**27**:4091-4093. DOI: 10.1039/B902132A

[60] Fritea L, Gorgy K, Goff AL, Audebert P, Galmiche L, Săndulescu R, et al. Fluorescent and redox tetrazine films by host-guest immobilization of tetrazine derivatives within poly (pyrrole-β-cyclodextrin) films. Journal of Electroanalytical Chemistry. 2016; **781**:36-40. DOI: 10.1016/j.jelechem.2016.07.010

Chapter 5

Cyclodextrins as Bricks for Tuning Polymer Properties

Ludmila Aricov, Anca Ruxandra Leonties, Iulia Matei and Gabriela Ioniță

Abstract

Cyclodextrins are natural cyclic oligosaccharides with a cone shape delimiting a hydrophobic cavity. The rims of cyclodextrins can be functionalized in order to improve their properties. Based on this, cyclodextrins can be linked to polymer chains, which further allows the tuning of the polymer properties. This review describes the methods of polymer functionalization with cyclodextrins and highlights the changes in the physicochemical properties of these materials. This chapter is focused on polymers in solution and in gel states. Cyclodextrin-based polymers are evaluated by various physicochemical methods, such as rheology, calorimetry, and spectroscopy (electron paramagnetic resonance, fluorescence, nuclear magnetic resonance (NMR), Fourier transform infrared (FT-IR), etc.). Both natural and synthetic polymers are considered in this chapter.

Keywords: cyclodextrin-based polymer, synthesis, rheological behavior, EPR spectroscopy

1. Introduction

Cyclodextrins (CDs) are natural compounds, obtained by the enzymatic modification of starch [1], that consist of at least six glucopyranose units linked by α-1,4 glycosidic bond. They have a truncated cone shape delimiting a relatively hydrophobic cavity and two polar rims bearing primary hydroxyl groups (the narrow rim) and secondary hydroxyl groups (the broader rim) [2]. The main feature of CDs is the formation of noncovalent complexes through host-guest interaction, which is, in fact, a summation of several noncovalent steps and overall represents an entropically driven process [3]. The host-guest interactions are determined by different factors, such as size, shape, charge, or polarity of the molecular actors involved [4]. This type of complexation has been extensively documented since the discovery of CDs, especially for low-molecular-weight compounds, by using a variety of physicochemical methods depending on the properties of the guest molecules [5–7]. These methods refer to nuclear magnetic resonance (NMR), UV-Vis, and fluorescence spectroscopies, as well as calorimetric methods. Although these molecules were reported at the end of the 19th century, the explosion of their applications started in the eighth decade of the 20th century [8]. CDs can be involved in the formation of noncovalent interactions with polymers, giving rise to special types of assemblies known as rotaxanes or pseudorotaxanes [9, 10].

The other feature of CDs is the presence of numerous primary and secondary hydroxyl groups that allow a facile derivatization in order to obtain new molecules that can be used as building blocks for large assemblies. Owing to the difference in reactivity of the primary and secondary groups, it is possible to control the functionalization, which ensures a selectivity of this process. The easiest way to synthesize monoderivatives is by obtaining monotosylated CD, especially for β-CD. This synthesis was studied in detail. A few syntheses are available in order to obtain pure CDs monotosylated at the primary hydroxyl rim that can be used further as bricks for preparing other derivatives. Monotosylated CD can be easily obtained in aqueous alkaline solution in good yields [11]. This derivative will be easily transformed further into amines [12] or thiols [13] that can be modified through maleimides or iodoacetamides [14]. There are also strategies describing the functionalization of primary and secondary rims that can allow obtaining of large supramolecular assemblies. It can be taken into account that the functionalization of the secondary rim is more difficult than the modification of the primary rim, as the hydroxyls that mark the larger rim require a strong base to become activated. In the review of Liu *et al.*, different ways to functionalize CDs at the secondary rims are described [15]. This method has been used to introduce sensing groups that allow studying supramolecular complexes of CDs by electron paramagnetic resonance (EPR) spectroscopy or fluorescence spectroscopy [14, 16]. The applications of these functionalization reactions will be referred in the cases discussed in this chapter.

In this review, we will focus on the two main features of CDs: to generate large supramolecular assemblies through host-guest interactions and to use functionalization in order to improve the properties of polymeric systems by the incorporation of host CD units. The polymeric systems taken into discussion refer to polyrotaxanes and pseudorotaxanes, the particular cases of sliding gels, the pluronic gels, and the functionalization of various synthetic and natural polymers with CD units. An important part will be focused on noncovalent and covalent hydrogels containing CD units.

2. Noncovalent interactions between polymers and cyclodextrins

Rotaxanes are supramolecular structures consisting of at least one ring threaded through an axial molecule, with the particularity that the dissociation of this assembly is hindered by bulky groups at the ends of the axial molecule [17]. CDs may play the role of a ring that is lined up on a polymer chain. The general procedures for the synthesis of polyrotaxanes involve two pathways: the threading followed by rotaxination and the slippage of CD over a bulky group (**Figure 1**). These methods are used for the preparation of CD-based pseudorotaxanes and rotaxanes in aqueous media [18]. The formation of pseudorotaxanes is dependent on the relation between the sizes of the CD cavity and polymeric chain. Thus, α-CD will generate polyrotaxanes and pseudopolyrotaxanes with polyethylene glycols and polyamines [19, 20], while β-CD with polypropylene glycols [21] and γ-CD with polyvinyl alcohol [22], which often leads to gel systems. The endcapping of polyethylene glycols with bulky substituents, such as bis(3,5-dinitrobenzoyl) and bis(2,4-dinitrobenzoyl), will favor the formation of polyrotaxanes with γ-CD [23].

Many water-soluble polymers act as nonionic surfactants by self-association in micelles that are sensitive to temperature and the presence of other molecules.

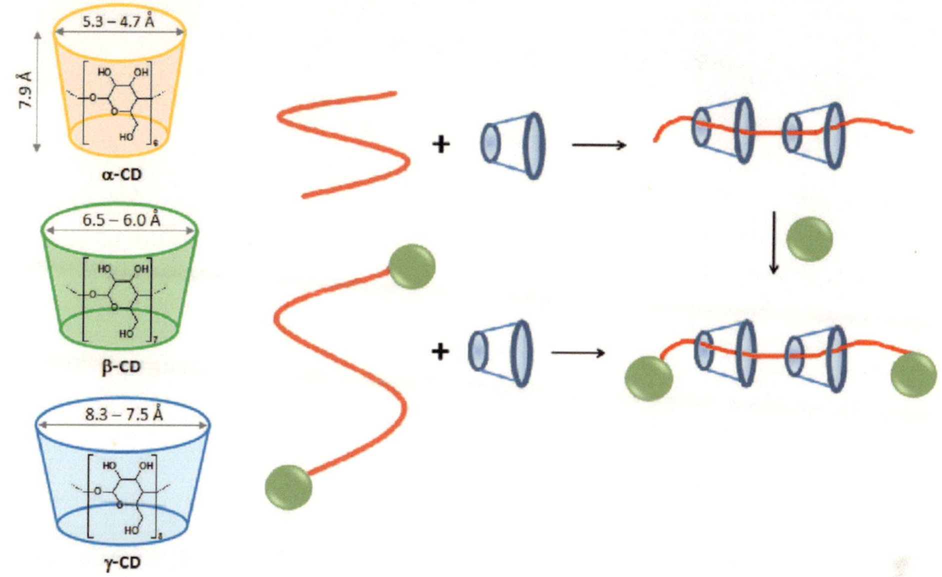

Figure 1.
Schematic representation of CDs and pseudopolyrotaxanes.

CDs can form inclusion complexes with nonionic surfactants from the Triton X or pluronic classes. In the case of Triton X, α-CD can complex the polyethylene chain, while β-CD complexes the nonpolar head (the iso-octylphenyl and phenyl groups) [24]. CDs can also modify the critical micelle concentration or are able to disrupt the micelles due to the complexation of polymers or surfactants [25]. As an example, the Triton X-100 nonionic surfactant, consisting of a short chain of polyethylene glycol and an aromatic nonpolar head, is able to form micelles at concentrations higher than 2.2 mM [26]. CDs can form inclusion complexes with this molecule with different geometries and stoichiometries depending on the CD size. Using isothermal titration calorimetry (ITC) measurements, it was possible to evaluate the binding constants for β-CD and γ-CD by assuming a stoichiometry of 1:1 and 1:2 for β-CD and of 1:1 for γ-CD, both involving the complexation of the nonpolar surfactant head. Conversely, in the case of α-CD, the complexation was supposed to be through the formation of pseudorotaxane by the inclusion of the polyethylene chain, assuming 1:5 stoichiometry, although the ITC data were not conclusive [27]. The interaction of β-CD with Triton X-114 led to the formation of larger aggregates involving hydrogen bonds with β-CD. This has been studied as a function of β-CD concentration and temperature, and it was observed that, at higher β-CD concentration, a transition from micelle to vesicle occurred. This effect is different from the more common effect of CDs on surfactant aggregation [28].

The effects of various CDs on the micellization and gelation of pluronics have also been reported. For such systems, changes in micellar concentration, in gelation, as well as changes of the hydration layer around the polymer chains during phase transition, when CDs are placed among polymeric chains, can be the result of pseudorotaxane formation [29–33]. Being water soluble, CDs will be more probable to target the region of the micelles placed at the water interface. Two studies involving EPR and fluorescence spectroscopies along with rheological and tube inversion methods explore the effect of 2-hydroxypropyl-β-CD on the micelle-to-gel phase transition of pluronic F127 [34, 35]. It was revealed that the spectral parameters of

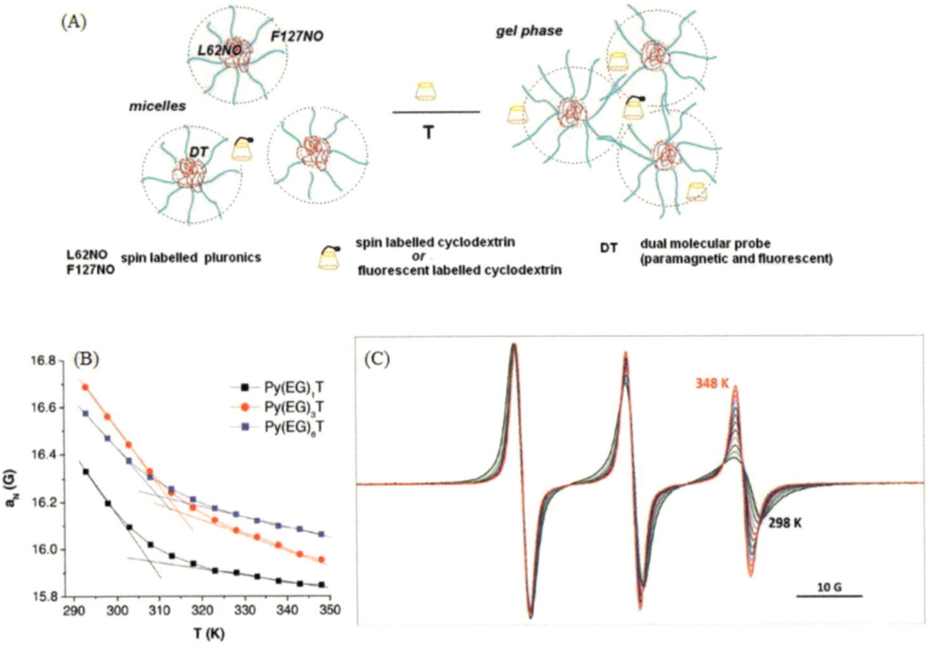

Figure 2.
Representation of the micelle-to-gel transition in F127/HPB systems (A), variation of hyperfine splitting constant with temperature (B) and variation of EPR spectra of spin probe with temperature (C) in F127 system [35].

molecular probes (commercially available spin probes, spin-labeled CDs, CDs labeled with fluorophores, or dual molecular probes) deviate from the linear dependence with temperature (**Figure 2**), thus indicating that the macroscopically observed phase transformation is related to changes at the nanoscale level. The results also led to the conclusion that the presence of CDs at the used concentrations does not induce micellar rupture but determines an increase in the micellar water content, which suggests an increase of the micellar size and excludes the formation of pseudopolyrotaxanes.

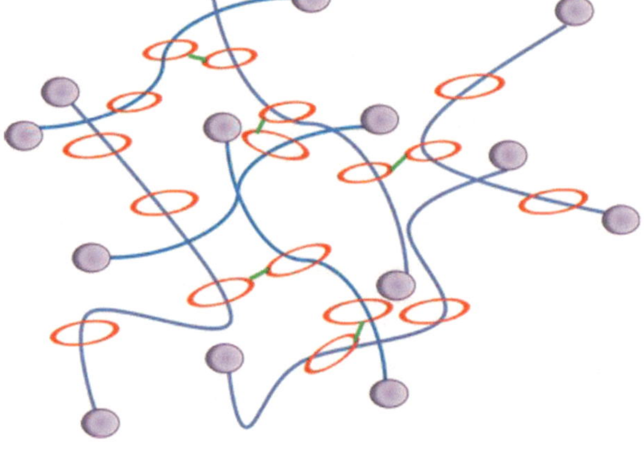

Figure 3.
Schematic representation of the supramolecular network of a sliding gel.

Rotaxanes and pseudorotaxanes generate networks that lead to the formation of hydrogels. A particular case refers to the formation of supramolecular networks named sliding gels that are characterized by mobile or sliding crossing points. A classical sliding network results from the intermolecular crosslinking of α-CD/polyethylene glycol pseudorotaxanes (**Figure 3**) [36–38]. The crossing points are, thus, mobile, and this will determine a mobility of the overall network. These gels are formed by linking CD units that belong to different chains.

The properties of these sliding gels can be described by topological parameters such as the complexation degree (number of CD units on a polymer chain), the crosslinker fraction (defined as the ratio of the mole number of crosslinker on the mole number of CDs), and the interactions between the swelling solvent and the constitutive parts of the network [39]. In many cases, the sliding motion depends on the swelling solvent, as well as on other factors such as the pH. For instance, in the case of pseudorotaxanes formed between a triblock copolymer consisting of polyethylene amine/polyethylene glycol/polyethylene amine, and α-CD, with crossing points obtained in the presence of 1,10-carbonyldiimidazole, the gel properties are dependent on the pH value [40].

3. Cyclodextrins covalently attached to polymeric chains

This section aims to summarize new research that has emerged in the past few years on materials composed of synthetic or natural polymers to which CD derivatives have been chemically attached. The polymers discussed here were chosen considering the extensive investigations and applications in pharmaceutics and biomedicine as drug excipients, biocompatible alternative materials in tissue engineering, contrast enhancers, molecular recognition models, etc. To achieve novel and/or enhanced properties of these classes of compounds, the functionalization with different CD units is employed. To covalently attach CDs onto polymer chains, multiple types of crosslinking agents (citric acid, epichlorohydrin, aldehydes, carbodiimides, and amines) can be used. In recent years, polymers functionalized with CD units have been studied for the development of a variety of polymeric networks [41, 42]. Moreover, the polymeric material based on CD units can be modulated in such a way to form nano/micro/macroparticles, gels, micelles, coating films, or fibers [43–45]. Many studies in this regard are performed using chitosan, mostly due to its remarkable biological properties, including antimicrobial activity, nontoxicity, biocompatibility, and biodegradability, and at the same time to the possibility to functionalize it [46, 47].

For instance, Campos *et al.* took this advantage and appended β-CD on chitosan nanoparticles in the presence of N-(3-dimethylaminopropyl)-N′-ethylcarbodiimide (EDC) and N-hydroxysuccinimide (NHS), thus developing a potential carrier for botanical pesticides [48]. A green-assembly strategy to obtain β-CD-chitosan functionalized graphene oxide hydrogels in the presence of sodium ascorbate has been reported, with application as recyclable decontaminants in wastewater treatment [49]. Both simple polymers and polymer mixtures are favorable to the attachment of CD units, as demonstrated by the research of Hardy *et al.* regarding the complex formed between chitosan and alginate for the release of piroxicam [50]. The research revealed that CD appending on chitosan generates a decrease in the number of amino groups, thus modulating the alginate complexation. In a recent study, alginate has also been functionalized with either host units (n-alkyl amine CD derivatives) or guest units (adamantane) [42]. In this particular case, the grafting procedure was performed in aqueous media in the presence of EDC and N-hydroxysulfosuccinimide (NHSS). The properties of the gel obtained by mixing the two types of alginates in the presence of Ca^{2+} ions were influenced by the host-guest interaction and the length of the alkyl chain of the β-CD derivatives. Thus, by mixing the decorated alginates with CD and adamantyl units, materials suitable for encapsulating both large molecules and small species can be obtained. In addition, the functionalization of alginate with units of β-CD derivatives by using Cu(I)-catalyzed azide-alkyne cycloaddition click reaction led to materials with good drug release properties [51]. The same procedure was used in another study, showing that the grafting degree of β-CD on alginate can be controlled and modulates the release/uptake of the model molecule methyl orange [52]. In addition, a composite based on sodium alginate grafted with β-CD using epichlorohydrin and NaOH was successfully obtained for the first time and used as a matrix in the immobilization of *Arthrobacter simplex* cells for cortisone acetate biotransformation [53].

Not only natural polymers were grafted with CD derivatives, but also synthetic ones. In this regard, the polyacrylic acid was modified by attaching a CD derivative (2-aminoethyl)amino-deoxy-β-CD) [54]. The study targeted to obtain the self-assembly of β-CD and adamantyl moieties covalently linked to polyacrylate networks for application in controlled complexation and release of ethyl orange, methyl orange, and methyl red. Another study developed a multistimulus responsive supramolecular hydrogel based on host-guest and electrostatic interactions between β-CD dimer and methoxy-azobenzene molecules grafted on polyacrylic acid [55]. The obtained materials showed thermo-, photo-, and pH-responsive behavior determining a reversible sol-gel transition.

Chabalala *et al.* described the grafting of β-CD molecules on polyacrylonitrile using citric and sulfuric acids as crosslinkers [56]. The nanofiber membranes produced by the electrospinning method were used for the adsorption of bromophenol blue and atrazine. Mono-(6-ethylenediamine-6-deoxy)-β-CD was appended in the presence of EDC/NHS crosslinking agents onto the external surface of a plasma separation membrane based on polyvinylidene fluoride [57].

4. Polymer gels containing cyclodextrins

Polymers containing CDs have the ability to form supramolecular hydrogels mostly due to host-guest complexation or by the inclusion of linear polymeric chains into host cavities [58]. The latter leads to the formation of pseudorotaxanes. Literature data show that, unlike β-CD, α and γ-CD are able to generate interactions favorable to the obtaining pseudorotaxanes [59]. However, in particular cases, pseudorotaxanes were prepared using β-CD and polylactic acid or polypropylene glycol [60, 61].

Several studies regarding host-guest interactions in hydrogel systems are reported in the literature [60, 62, 63]. By appending host and guest units to the polymer chains, one can modify and control their behavior in solutions and obtain gel systems. This strategy ensures conditions for the creation of topological crosslinks based on host-guest interactions, which have the advantage of being reversible and movable. Host and guest units can be grafted on the chains of numerous polymers (hyaluronic acid [64, 65], carboxymethyl cellulose [66, 67], sodium alginate [42, 68], polyacrylic acid [55, 69, 70], polyvinyl alcohol [71, 72], polymethyl vinyl ether-alt-maleic acid [73, 74], poly-N-isopropylacrylamide [75], and polyethylene glycol [76]). These supramolecular gels find use in the medical field for drug delivery, tissue culture, and medical treatments [77, 78]. Depending on the desired application, hydrogels formed by host-guest interactions can be generated in several ways. Therefore, CD having the role of guest molecules can be grafted separately on polymer chains and mixed with polymers bearing guest grafts to form supramolecular assemblies [41, 60].

The formation of alginate gels in the presence of divalent cations was monitored by EPR spectroscopy considering the changes in the dynamics of spin-labeled alginate chain [79]. In a recent study, we showed that the functionalization of alginate with CD (as host) or adamantane (as guest) influences the properties of ionotropic generated gel in the presence of Ca^{2+} ions [42]. As a consequence, polymer functionalization and subsequent interactions between the appended host and guest units change the morphology of the resulting xerogels (**Figure 4**).

In fact, the derivatization, together with the host-guest interactions, has an impact on the rheological properties, i.e., the hydrogels made from a mixture of adamantane-functionalized and CD-functionalized alginates presented higher storage and elastic

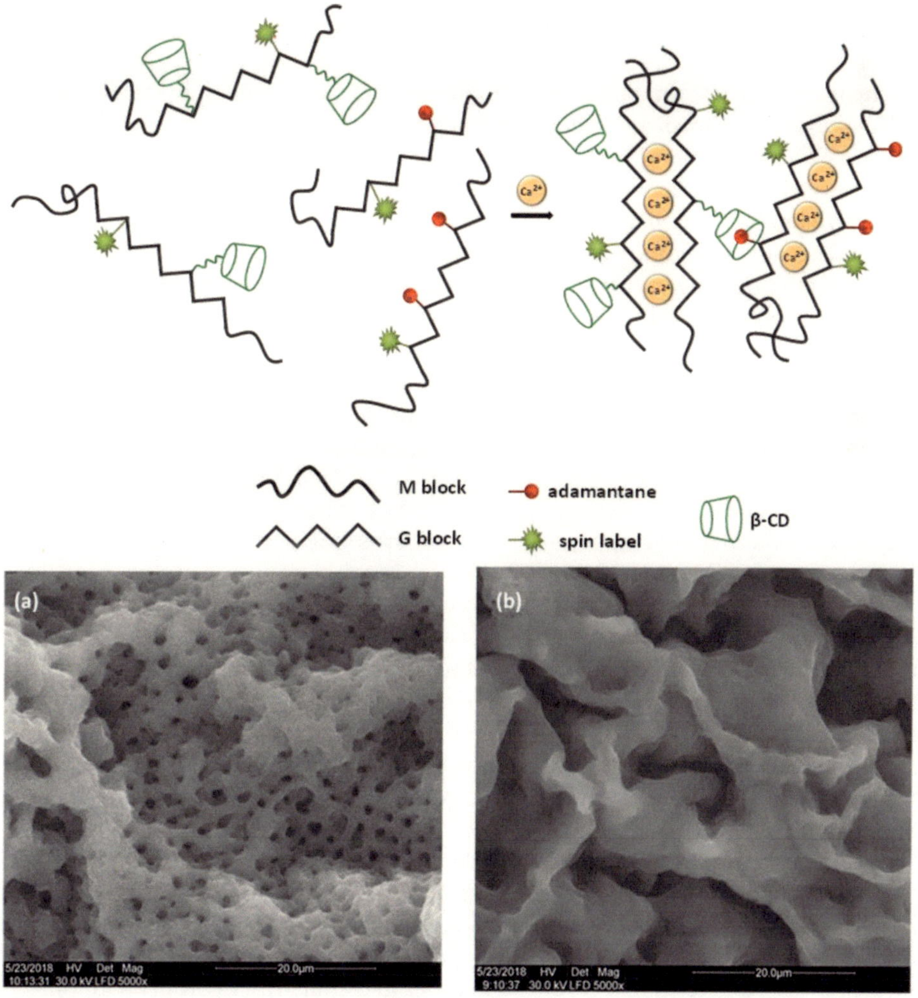

Figure 4.
Schematic representation of ionotropic gelation of functionalized alginates. SEM images of alginate xerogels: (a) nonfunctionalized alginate and (b) alginate functionalized with CD and adamantane units [42].

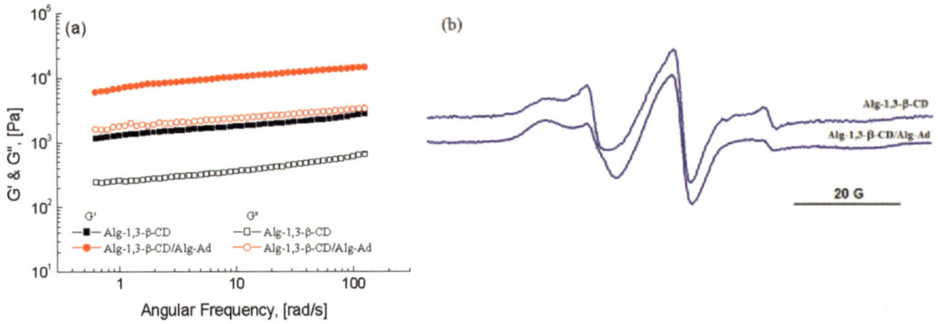

Figure 5.
(a) Viscoelastic properties and (b) EPR spectra of alginate hydrogels functionalized with β-CD or adamantane units [42].

moduli values (**Figure 5a**). Appending the paramagnetic moieties to these functionalized alginates allows evidencing the changes at the molecular level of the dynamic of the overall motion (**Figure 5b**). All these findings led to the conclusion that the presence of host-guest interactions can modulate the features of alginate hydrogels.

The polymerization of the inclusion complex is another technique that can be used to obtain CD-based hydrogels. Ikura *et al.* approached the free-radical copolymerization of polyethyl acrylate crosslinked with peracetylated γ-cyclodextrin methylacrylamide monomer and acrylate monomers and investigated the effect of the size of the main chain monomers on the formation of movable crosslinking points [80]. Thus, it was observed that the small polymer main chains penetrated the CD units and acquired the role of movable crosslinking points in the hydrogel, whereas copolymerization with bulky monomers leads to hydrogels without movable crosslinking points.

In aqueous media, host-capped polymers can be mixed with guest-grafted chains to obtain hydrogels. Ioniță *et al.* showed that the reaction of isocyanate end-capped polyethylene glycol with β-CD leads to the formation of a fibrous gel with covalent network [81]. Moreover, by adding spin probes (e.g., TEMPO and adamantane-TEMPO), it was possible to determine the extent by which gel fibers were affected by hydrogen bonding interactions with solvent, crosslinks density, or temperature. The study revealed that, at low temperature, ice crystallization is prevented inside the gels, and this phenomenon is accompanied by the formation of supercooled water.

Another method to obtain hydrogels is the mixing of end-capped guest crosslinkers with certain host-grafted polymers [41, 60]. A recent study has shown that the multivalence effect within a polyethylene glycol-adamantane/β-CD-alginate system can be quantified to create hydrogel-like cell matrices [82]. The complexation of CD functionalized alginate with adamantyl end groups on polyethylene glycol chains changed the valence of the system. Thus, a correlation could be observed between the multivalence generated by the variation in the number of polyethylene glycol arms and the strength of binding affinities inside the hydrogel.

A particular type of gel is formed between guest-grafted polymers that are capable of forming multilayer vesicles and small host molecules grafted on different polymeric chains. A good example is the gel formed by thiolated monolith polymers, in which β-CD vesicles were introduced in order to formulate a hydrogel with pH-responsive properties [83].

Polymeric gel formation can be described using rheological, viscosity, and dynamic light scattering measurements. These methods provide global information on such systems. Spectroscopic methods, such as fluorescence, IR, or UV-Vis spectroscopy, are often used to describe changes in the organization of macromolecules that are usually governed by noncovalent interactions [84]. In the particular case of gel formation, electron microscopy techniques are used to evidence the gel fibers. An interesting, powerful, but still rarely used approach in studying gels involves EPR spectroscopy [34, 79, 81, 84, 85]. The EPR spectroscopy is suitable to study polymer systems and gels as the method can provide insights into local, static, and dynamic properties of these systems. This method can evidence nanoscale inhomogeneities in polymers systems [86, 87].

By using spin-labeled CDs, it was possible to monitor the gel formation process, while the diffusion of various spin probes can evidence the nonuniform properties of covalent gels [81, 88, 89]. Other EPR studies explored the self-assembly of pluronic F127 leading to gel phase as a function of temperature and concentration of CD [34, 35] or the formation of supramolecular gels resulted by the assembly of low-molecular-weight gelators [85].

5. Applications

Over the past four years, data from the literature indicate some comprehensive reviews of CD-based polymer applications [63, 90–93]. The most common applications of these materials are in the field of drug delivery [60, 94, 95], biomedical engineering [90, 96, 97], food industry [93, 98, 99], responsive adhesives [100, 101], coatings [102, 103], sensors [104, 105], and environmental remediation [106, 107].

Table 1 exemplifies some of the most recent applications reported in the literature on polymers functionalized with CD units.

Material	Applications	Ref.
β-CD-chitosan nanoparticles	Carrier for carvacrol and linalool	[48]
β-CD-chitosan/graphene oxide hydrogel	Removal of methylene blue	[49]
β-CD-chitosan/alginate complex	Encapsulation and release of piroxicam	[50]
β-CD-alginate hydrogels	Paclitaxel drug release	[51]
β-CD-alginate gel beads	Release of methyl orange	[52]
β-CD-alginate	Matrix for *Arthrobacter simplex* cell immobilization	[53]
β-CD-polyacrylonitrile nanofiber	Adsorption of bromophenol blue and atrazine from aqueous systems	[56]
β-CD polymer-tetrafluoroterephthalonitrile	Removal of malachite green	[108]
β-CD polymer-tetrafluoroterephthalonitrile	Agent for monitoring endocrine disrupting chemicals from water	[109]
β-CD based polymeric adsorbent	Pollutants removal (rhodamine B, Congo red and cadmium ions) from wastewater	[110]
2-hydroxypropyl-β-CD-polyacrylic acid	Removal of ibuprofen	[111]
β-CD-activated charcoal-Na alginate magnetic beads	Removal of methyl violet, brilliant green, norfloxacin, ciprofloxacin and copper ions from aqueous systems	[112]

Material	Applications	Ref.
β-CD-carboxymethyl chitosan	Sensor for direct determination of manganese	[113]
β-CD-alginate-graphene oxide hydrogel	Injectable hydrogel for soft tissue engineering	[114]
β-CD-epichlorohydrin-carboxymethyl chitosan	Bioactive enhancement of cyanidin-3-glucoside	[115]
β-CD-polyurethane/chitosan	Gentamicin sulphate drug release	[116]
β-CD-Fe_3O_4-chitosan nanoparticles	Support for lipase immobilization	[117]
β-CD-chitosan beads	Support for keratinase immobilization	[118]

Table 1.
Applications of polymers functionalized with β-cyclodextrin.

6. Conclusions and perspectives

The general lines of polymer functionalization with cyclodextrins and resulting changes of their properties have been reviewed in this chapter. These can be applied for generating and studying other polymer-cyclodextrin systems. The EPR spectroscopy can be used as a method for proving changes at nanoscale level in such systems due to host-guest interactions occurring in polymer systems. The EPR data can be linked with other experimental data provided by classical methods used to characterize polymers in solution and in gel states, like rheology, electron microscopy techniques, etc.

Acknowledgements

The chapter was supported by the Romanian National Authority for Scientific Research, CNCS-UEFISCDI, project code PN-III-P1-1.1-PD-2021-0169 and contract number PD 106 and Romanian Academy within the research programs "*Colloids and dispersed systems*" and "*EPR and fluorescence studies on supramolecular interactions in inhomogeneous systems*" of the "Ilie Murgulescu" Institute of Physical Chemistry.

Conflict of interest

The authors declare no conflict of interest.

Author details

Ludmila Aricov*, Anca Ruxandra Leontieș, Iulia Matei and Gabriela Ioniță
"Ilie Murgulescu" Institute of Physical Chemistry of the Romanian Academy, Bucharest, Romania

*Address all correspondence to: aricov_ludmila@yahoo.com

References

[1] Biwer A, Antranikian G, Heinzle E. Enzymatic production of cyclodextrins. Applied Microbiology and Biotechnology. 2002;**59**:609-617. DOI: 10.1007/s00253-002-1057-x

[2] Szejtli J. Introduction and general overview of cyclodextrin. Chemical Reviews. 1998;**98**:1743-1753. DOI: 10.1021/cr970022c

[3] Pluth MD, Raymond KN. Reversible guest exchange mechanisms in supramolecular host-guest assemblies. Chemical Society Reviews. 2007;**36**: 161-171. DOI: 10.1039/b603168b

[4] Krieg E, Bastings MMC, Besenius P, Rybtchinski B. Supramolecular polymers in aqueous media. Chemical Reviews. 2016;**116**:2414-2477. DOI: 10.1021/acs.chemrev.5b00369

[5] Muankaew C, Loftsson T. Cyclodextrin-based formulations: A non-invasive platform for targeted drug delivery. Basic & Clinical Pharmacology & Toxicology. 2018;**122**:46-55. DOI: 10.1111/bcpt.12917

[6] He J, Li Y, Wang C, Zhang K, Lin D, Kong L, et al. Rapid adsorption of Pb, Cu and Cd from aqueous solutions by β-cyclodextrin polymers. Applied Surface Science. 2017;**426**:29-39. DOI: 10.1016/j.apsusc.2017.07.103

[7] Cid-Samamed A, Rakmai J, Mejuto JC, Simal-Gandara J, Astray G. Cyclodextrins inclusion complex: Preparation methods, analytical techniques and food industry applications. Food Chemistry. 2022;**384**:132467. DOI: 10.1016/j.foodchem.2022.132467

[8] Crini G. Review: A history of cyclodextrins. Chemical Reviews. 2014;**114**:10940-10975. DOI: 10.1021/cr500081p

[9] Akae Y, Sogawa H, Takata T. Synthesis of a structure-definite α-cyclodextrin-based macromolecular [3]Rotaxane using a size-complementary method. Angewandte Chemie, International Edition. 2018;**57**:11742-11746. DOI: 10.1002/anie.201807261

[10] Teuchert C, Michel C, Hausen F, Park D-Y, Beckham HW, Wenz G. Cylindrical polymer brushes by atom transfer radical polymerization from cyclodextrin–PEG polyrotaxanes: Synthesis and mechanical stability. Macromolecules. 2013;**46**:2-7. DOI: 10.1021/ma302204a

[11] Byun SH, Zhong N, Bittman R. 6A-O-p-Toluenesulfonyl-b-cyclodextrin. Organic Syntheses. 2000;**77**:225-230. DOI: 10.15227/orgsyn.077.0225

[12] Ohashi H, Hiraoka Y, Yamaguchi T. An autonomous phase transition-complexation/decomplexation polymer system with a molecular recognition property. Macromolecules. 2006;**39**:2614-2620. DOI: 10.1021/ma052509q

[13] Fujita K, Ueda T, Imoto T, Tabushi I, Toh N, Koga T. Guest-induced conformational change of β-cyclodextrin capped with an environmentally sensitive chromophore. Bioorganic Chemistry. 1982;**11**:72-84. DOI: 10.1016/0045-2068(82)90049-9

[14] Ionita G, Chechik V. Spin-labelled cyclodextrins as hosts for large supramolecular assemblies. Organic & Biomolecular Chemistry. 2005;**3**:3096-3098. DOI: 10.1039/b508256k

[15] Liu J, Yu P, Sollogoub M, Zhang Y. Functionalized cyclodextrins and their

applications in biodelivery. In: Liu Y, Chen Y, Zhang HY, editors. Handbook of Macrocyclic Supramolecular Assembly. Singapore: Springer; 2019. pp. 1-39. DOI: 10.1007/978-981-13-1744-6_15-1

[16] Wang Y, Ikeda T, Ikeda H, Ueno A, Toda F. Dansyl- β -cyclodextrins as fluorescent sensors responsive to organic compounds. Bulletin Chemical Society of Japan. 1994;**67**:1598-1607. DOI: 10.1246/bcsj.67.1598

[17] Wenz G, Han BH, Müller A. Cyclodextrin rotaxanes and polyrotaxanes. Chemical Reviews. 2006;**782**:817. DOI: 10.1021/cr970027+

[18] Harada A, Hashidzume A, Yamaguchi H, Takashima Y. Polymeric rotaxanes. Chemical Reviews. 2009;**109**:5974-6023. DOI: 10.1021/cr9000622

[19] Yamada S, Sanada Y, Tamura A, Yui N, Sakurai K. Chain architecture and flexibility of α-cyclodextrin/PEG polyrotaxanes in dilute solutions. Polymer Journal. 2015;**47**:464-467. DOI: 10.1038/pj.2015.18

[20] Huang F, Gibson HW. Polypseudorotaxanes and polyrotaxanes. Progress in Polymer Science. 2005;**30**:982-1018. DOI: 10.1016/j.progpolymsci.2005.07.003

[21] Okada M, Harada A. Preparation of β-cyclodextrin polyrotaxane: Photodimerization of pseudo-Polyrotaxane with 2-Anthryl and triphenylmethyl groups at the ends of poly(propylene glycol). Organic Letters. 2004;**6**:361-364. DOI: 10.1021/ol0361608

[22] Hernández R, Rusa M, Rusa CC, López D, Mijangos C, Tonelli AE. Controlling PVA hydrogels with γ-cyclodextrin. Macromolecules. 2004;**37**:9620-9625. DOI: 10.1021/ma048375i

[23] Harada A, Li J, Kamachi M. Double-stranded inclusion complexes of cyclodextrin threaded on poly(ethylene glycol). Nature. 1994;**126**:128. DOI: 10.1038/370126a0

[24] Harada A, Okumura H, Okada M, Suzuki S, Kamachi M. Site-selective complexation of amphiphilic compounds by cyclodextrins. Chemistry Letters. 2000;**29**:548-549. DOI: 10.1246/cl.2000.548

[25] Kang Y, Guo K, Li BJ, Zhang S. Nanoassemblies driven by cyclodextrin-based inclusion complexation. Chemical Communications. 2014;**250**:11083-11109. DOI: 10.1039/c4cc03131h

[26] Tiller GE, Mueller TJ, Dockter ME, Struve WG. Hydrogenation of Triton X-100 eliminates its fluorescence and ultraviolet light absorption while preserving its detergent properties. Analytical Biochemistry. 1984;**141**:262-266. DOI: 10.1016/0003-2697(84)90455-X

[27] Müller BK, Ritter H. Scrutinizing ITC-study on the formation of inclusion complexes of nonionic surfactant Triton X-100 and cyclodextrins. Journal of Inclusion Phenomena and Macrocyclic Chemistry. 2012;**72**:157-164. DOI: 10.1007/s10847-011-9955-0

[28] Xu HN, Ma SF, Chen W. Unique role of β-cyclodextrin in modifying aggregation of Triton X-114 in aqueous solutions. Soft Matter. 2012;**8**:3856-3863. DOI: 10.1039/c2sm07371d

[29] Valero M, Grillo I, Dreiss CA. Rupture of pluronic micelles by Di-methylated β-cyclodextrin is not due to polypseudorotaxane formation. The Journal of Physical Chemistry. B. 2012;**116**:1273-1281. DOI: 10.1021/jp210439n

[30] Joseph J, Dreiss CA, Cosgrove T, Pedersen JS. Rupturing polymeric micelles with cyclodextrins. Langmuir. 2007;**23**:460-466. DOI: 10.1021/la061850g

[31] Dreiss CA, Nwabunwanne E, Liu R, Brooks NJ. Assembling and de-assembling micelles: Competitive interactions of cyclodextrins and drugs with Pluronics. Soft Matter. 2009;**5**:1888-1896. DOI: 10.1039/b812805g

[32] Valero M, Dreiss CA. Growth, shrinking, and breaking of pluronic micelles in the presence of drugs and/or β-cyclodextrin, a study by small-angle neutron scattering and fluorescence spectroscopy. Langmuir. 2010;**26**:10561-10571. DOI: 10.1021/la100596q

[33] Valero M, Hu W, Houston JE, Dreiss CA. Solubilisation of salicylate in F127 micelles: Effect of pH and temperature on morphology and interactions with cyclodextrin. Journal of Molecular Liquids. 2021;**322**:114892. DOI: 10.1016/j.molliq.2020.114892

[34] Micutz M, Matalon E, Staicu T, Angelescu D, Ariciu AM, Turcu IM, et al. The influence of hydroxy propyl β-cyclodextrin on the micellar to gel transition in F127 solutions investigated at macro and nanoscale levels. New Journal of Chemistry. 2014;**38**:2810-2812. DOI: 10.1039/c4nj00123k

[35] Baratoiu R, Mocanu S, Matei I, Bem M, Hristea E, Tecuceanu V, et al. A comparison of the behavior of monomolecular and dual molecular probes in F127/cyclodextrin systems. Macromolecular Chemistry and Physics. 2019;**220**:1800489-1800497. DOI: 10.1002/macp.201800489

[36] Okumura Y, Ito K. The polyrotaxane gel: A topological gel by figure-of-eight cross-links. Advanced Materials. 2001;**13**:485-487. DOI: 10.1002/1521-4095(200104)13:7<485::AID-ADMA485>3.0.CO;2-T

[37] Harada A, Li J, Kamachi M. Preparation and Properties of Inclusion Complexes of Poly(ethylene glycol) with α-Cyclodextrin. Macromolecules. 1993;**26**:5698-5703. DOI: 10.1021/ma00073a026

[38] Fleury G, Schlatter G, Brochon C, Hadziioannou G. From high molecular weight precursor polyrotaxanes to supramolecular sliding networks. Polymer. 2005;**46**:8494-8501. DOI: 10.1016/j.polymer.2005.02.125

[39] Fleury G, Schlatter G, Brochon C, Travelet C, Lapp A, Lindner P, et al. Topological polymer networks with sliding cross-link points: The "Sliding gels". Relationship between their molecular structure and the viscoelastic as well as the swelling properties. Macromolecules. 2007;**40**:535-543. DOI: 10.1021/ma0605043

[40] Karaky K, Brochon C, Schlatter G, Hadziioannou G. pH-Switchable supramolecular "sliding" gels based on polyrotaxanes of polyethyleneimine-block-poly(ethylene oxide)-block-polyethyleneimine block copolymer and α-cyclodextrin: Synthesis and swelling behaviour. Soft Matter. 2008;**4**:1165-1168. DOI: 10.1039/b803670e

[41] Arslan M, Sanyal R, Sanyal A. Cyclodextrin embedded covalently crosslinked networks: Synthesis and applications of hydrogels with nano-containers. Polymer Chemistry. 2020;**11**:615-629. DOI: 10.1039/C9PY01679A

[42] Popescu EI, Aricov L, Mocanu S, Matei I, Hristea E, Baratoiu R, et al. Subtle influence on alginate gel properties through host–guest interactions between

covalently appended cyclodextrin and adamantane units. New Journal of Chemistry. 2021;**145**:8083-8089. DOI: 10.1039/D1NJ01278A

[43] Lu J, Li X, Qiu C, McClements DJ, Jiao A, Wang J, et al. Preparation and characterization of food-grade pickering emulsions stabilized with chitosan-phytic acid-cyclodextrin nanoparticles. Food. 2022;**11**:450. DOI: 10.3390/foods11030450

[44] Adeli F, Abbasi F, Babazadeh M, Davaran S. Thermo/pH dual-responsive micelles based on the host–guest interaction between benzimidazole-terminated graft copolymer and β-cyclodextrin-functionalized star block copolymer for smart drug delivery. Journal of Nanobiotechnology. 2022;**20**:1-20. DOI: 10.1186/s12951-022-01290-3

[45] Li X, Li C, Goh K, Chong TH, Wang R. Layer-by-layer aided β-cyclodextrin nanofilm for precise organic solvent nanofiltration. Journal of Membrane Science. 2022;**652**:120466. DOI: 10.1016/j.memsci.2022.120466

[46] Verma M, Lee I, Hong Y, Kumar V, Kim H. Multifunctional β-Cyclodextrin-EDTA-Chitosan polymer adsorbent synthesis for simultaneous removal of heavy metals and organic dyes from wastewater. Environmental Pollution. 2022;**292**:118447. DOI: 10.1016/j.envpol.2021.118447

[47] Jarupatnadech T, Chalitangkoon J, Monvisade P. Colorimetric oxygen indicator films based on β-cyclodextrin grafted chitosan/montmorillonite with redox system for intelligent food packaging. Packaging Technology and Science. 2022;**35**:515-525. DOI: 10.1002/pts.2648

[48] Campos EVR, Proença PLF, Oliveira JL, Melville CC, Vechia JFD, de Andrade DJL, et al. Chitosan nanoparticles functionalized with β-cyclodextrin: A promising carrier for botanical pesticides. Scientific Reports. 2018;**8**:2067. DOI: 10.1038/s41598-018-20602-y

[49] Liu Y, Huang S, Zhao X, Zhang Y. Fabrication of three-dimensional porous β-cyclodextrin/chitosan functionalized graphene oxide hydrogel for methylene blue removal from aqueous solution. Colloids and Surfaces A: Physicochemical and Engineering Aspects. 2018;**539**:1-10. DOI: 10.1016/j.colsurfa.2017.11.066

[50] Hardy A, Seguin C, Brion A, Lavalle P, Schaaf P, Fournel S, et al. β-Cyclodextrin-Functionalized chitosan/clginate compact polyelectrolyte complexes (CoPECs) as functional biomaterials with anti-inflammatory properties. ACS Applied Materials & Interfaces. 2018;**10**:29347-29356. DOI: 10.1021/acsami.8b09733

[51] Omtvedt LA, Kristiansen KA, Strand WI, Aachmann FL, Strand BL, Zaytseva-Zotova DS. Alginate hydrogels functionalized with β-cyclodextrin as a local paclitaxel delivery system. Journal of Biomedial Materials Research Part A. 2021;**109**:2625-2639. DOI: 10.1002/jbm.a.37255

[52] Omtvedt LA, Dalheim MØ, Nielsen TT, Larsen KL, Strand BL, Aachmann FL. Efficient grafting of cyclodextrin to alginate and performance of the hydrogel for release of model drug. Scientific Reports. 2019;**9**:9325. DOI: 10.1038/s41598-019-45761-4

[53] Shen Y, Niu L, Yu Z, Wang M, Shang Z, Yang Y. Sodium alginate-grafted β-cyclodextrins as a matrix for immobilized Arthrobacter simplex for cortisone acetate biotransfromation. Applied Surface Science. 2018;**444**:42-47. DOI: 10.1016/j.apsusc.2018.03.028

[54] Yan L, Pham D-T, Clements P, Lincoln FS, Wang J, Guo X, et al. β-Cyclodextrin- and adamantyl-substituted poly(acrylate) self-assembling aqueous networks designed for controlled complexation and release of small molecules. Beilstein Journal of Organic Chemistry. 2017;**13**:1879-1892. DOI: 10.3762/bjoc.13.183

[55] Yan H, Jiang Q, Wang J, Cao S, Qiu Y, Wang H, et al. A triple-stimuli responsive supramolecular hydrogel based on methoxy-azobenzene-grafted poly(acrylic acid) and β-cyclodextrin dimer. Polymer. 2021;**221**:123617. DOI: 10.1016/j.polymer.2021.123617

[56] Chabalala MB, Al-Abri MZ, Mamba BB, Nxumalo EN. Mechanistic aspects for the enhanced adsorption of bromophenol blue and atrazine over cyclodextrin modified polyacrylonitrile nanofiber membranes. Chemical Engineering Research and Design. 2021;**169**:19-32. DOI: 10.1016/j.cherd.2021.02.010

[57] Liu J, Lu X, Shu G, Ni F, Li K, Kong X, et al. Structure design and performance study on filtration-adsorption bifunctional blood purification membrane. Journal of Membrane Science. 2021;**636**:119535. DOI: 10.1016/j.memsci.2021.119535

[58] Domiński A, Konieczny T, Kurcok P. α-Cyclodextrin-based polypseudorotaxane hydrogels. Materials (Basel). 2019;**13**:133. DOI: 10.3390/ma13010133

[59] Hane FT, Fernando A, Prete BRJ, Peloquin B, Karas S, Chaudhuri S, et al. Cyclodextrin-Based Pseudorotaxanes: Easily conjugatable scaffolds for synthesizing hyperpolarized Xenon-129 magnetic resonance imaging agents. ACS Omega. 2018;**3**:677-681. DOI: 10.1021/acsomega.7b01744

[60] Fang G, Yang X, Chen S, Wang Q, Zhang A, Tang B. Cyclodextrin-based host–guest supramolecular hydrogels for local drug delivery. Coordination Chemistry Reviews. 2022;**454**:214352. DOI: 10.1016/j.ccr.2021.214352

[61] Brack W, Aissa AS, Backhaus T, Dulio V, Escher BI, Faust M, et al. Effect-based methods are key. The European Collaborative Project SOLUTIONS recommends integrating effect-based methods for diagnosis and monitoring of water quality. Environmental Sciences Europe. 2019;**10**. DOI: 10.1186/s12302-019-0192-2

[62] Liu G, Yuan Q, Hollett G, Zhao W, Kang Y, Wu J. Cyclodextrin-based host-guest supramolecular hydrogel and its application in biomedical fields. Polymer Chemistry. 2018;**9**:3436-3449. DOI: 10.1039/C8PY00730F

[63] Seidi F, Jin Y, Xiao H. Polycyclodextrins: Synthesis, functionalization, and applications. Carbohydrate Polymers. 2020;**242**:116277-116242. DOI: 10.1016/j.carbpol.2020.116277

[64] Yu B, Zhan A, Liu Q, Ye H, Huanga X, Shu Y, et al. A designed supramolecular cross-linking hydrogel for the direct, convenient, and efficient administration of hydrophobic drugs. International Journal of Pharmaceutics. 2020;**578**:119075. DOI: 10.1016/j.ijpharm.2020.119075

[65] Wu K, Wu X, Zhang Y, Chen S, Qiao Z, Wei D, et al. Semiconvertible hyaluronic hydrogel enabled red-light-responsive reversible mechanics, adhesion, and self-healing. Biomacromolecules. 2022;**23**:1030-1040. DOI: 10.1021/acs.biomac.1c01395

[66] Jiang X, Zeng F, Yang X, Yang X, Jian C, Zhang L, et al. Injectable self-healing cellulose hydrogel based

on host-guest interactions and acylhydrazone bonds for sustained cancer therapy. Acta Biomaterialia. 2022;**141**:102-113. DOI: 10.1016/j.actbio.2021.12.036

[67] Yao Y, Yu S, Shen Y, Wu H. Facile synthesis of self-dispersed β-cyclodextrin-coupled cellulose microgel for sustained release of vanillin. International Journal of Biological Macromolecules. 2022;**79**:208-270. DOI: 10.1016/j.ijbiomac.2022.03.071

[68] Yang S, Qin W, Zhao X, He F, Gong H, Liu Y, et al. Interfacial self-assembled behavior of pH/light-responsive host-guest alginate-based supra-amphiphiles for controlling emulsifying property. Carbohydrate Polymers. 2021;**266**:118121. DOI: 10.1016/j.carbpol.2021.118121

[69] Ding L, Xiang C, Zhou G. Silica nanoparticles coated by poly(acrylic acid) brushes via host-guest interactions for detecting DNA sequence of Hepatitis B virus. Talanta. 2018;**181**:72-181. DOI: 10.1016/j.talanta.2017.12.061

[70] Hou N, Wang R, Wang F, Bai J, Jiao T, Bai Z, et al. Self-assembled hydrogels constructed via host-guest polymers with highly efficient dye removal capability for wastewater treatment. Colloids and Surfaces A: Physicochemical and Engineering Aspects. 2019;**579**:123670. DOI: 10.1016/j.colsurfa.2019.123670

[71] Jia YG, Jin J, Liu S, Ren L, Luo J, Zhu XX. Self-healing hydrogels of low molecular weight poly(vinyl alcohol) assembled by host–guest recognition. Biomacromolecules. 2018;**19**:626-632. DOI: 10.1021/acs.biomac.7b01707

[72] Li JM, Hu CS, Shao JM, Li HJ, Li PY, Li XC, et al. Fabricating ternary hydrogels of P(AM-co-DMAEMA)/ PVA/ β-CD based on multiple physical crosslinkage. Polymer. 2017;**119**:152-159. DOI: 10.1016/j.polymer.2017.05.031

[73] Ren P, Wang F, Bernaerts KV, Fu Y, Hu W, Zhou N, et al. Self-assembled supramolecular hybrid hydrogels based on host–guest interaction: Formation and application in 3D cell culture. ACS Applied Bio Materials. 2020;**3**:6778. DOI: 10.1021/acsabm.0c00711

[74] Ma X, Zhou N, Zhang T, Hu W, Gu N. Self-healing pH-sensitive poly[(methyl vinyl ether)-alt-(maleic acid)]-based supramolecular hydrogels formed by inclusion complexation between cyclodextrin and adamantane. Materials Science and Engineering: C. 2017;**73**:357-365. DOI: 10.1016/j.msec.2016.12.039

[75] Zhou Z, Li G, Wang N, Guo F, Guo L, Liu X. Synthesis of temperature/pH dual-sensitive supramolecular micelles from β-cyclodextrin-poly(N-isopropylacrylamide) star polymer for drug delivery. Colloids and Surfaces. B, Biointerfaces. 2018;**172**:136-142. DOI: 10.1016/j.colsurfb.2018.08.031

[76] Wang J, Zhang X, Zhang S, Kang J, Guo Z, Feng B, et al. Semi-convertible hydrogel enabled photoresponsive lubrication. Matter. 2021;**4**:675-687. DOI: 10.1016/j.matt.2020.11.018

[77] Auzély-Velty R. Self-assembling polysaccharide systems based on cyclodextrin complexation: Synthesis, properties and potential applications in the biomaterials field. Comptes Rendus Chimie. 2011;**14**:166-177. DOI: 10.1016/j.crci.2010.04.019

[78] Skopinska-Wisniewska J, De la Flor S, Kozlowska J. From supramolecular hydrogels to multifunctional carriers for biologically active substances. International Journal of Molecular Sciences. 2021;**22**:7402. DOI: 10.3390/ijms22147402

[79] Ionita G, Ariciu AM, Smith DK, Chechik V. Ion exchange in alginate gels – dynamic behaviour revealed by electron paramagnetic resonance. Soft Matter. 2015;**11**:8968-8974. DOI: 10.1039/C5SM02062J

[80] Ikura R, Park J, Osaki M, Yamaguchi H, Harada A, Takashima Y. Supramolecular elastomers with movable cross-linkers showing high fracture energy based on stress dispersion. Macromolecules. 2019;**52**:6953-6962. DOI: 10.1021/acs.macromol.9b01198

[81] Ionita G, Ariciu AM, Turcu IM, Chechik V. Properties of polyethylene glycol/cyclodextrin hydrogels revealed by spin probes and spin labelling methods. Soft Matter. 2014;**10**:1778-1783. DOI: 10.1039/c3sm52004h

[82] Ooi HW, Kocken JMM, Morgan FLC, Malheiro A, Zoetebier B, Karperien M, et al. Multivalency enables dynamic supramolecular host–guest hydrogel formation. Biomacromolecules. 2020;**21**:2208-2217. DOI: 10.1021/acs.biomac.0c00148

[83] Zheng H, Li X, Jia Q. Design of pH-responsive polymer monolith based on cyclodextrin vesicle for capture and release of myoglobin. ACS Applied Materials & Interfaces. 2018;**10**:5909-5917. DOI: 10.1021/acsami.7b18999

[84] Ionita G. Characterization and tailoring the properties of hydrogels using spectroscopic methods. In: Majee SB, editor. Emerging Concepts in Analysis and Applications of Hydrogels. London: InTech; 2016. pp. 39-68. DOI: 10.5772/62900

[85] Ariciu AM, Staicu T, Micutz M, Neacsu MV, Ionita P, Tecuceanu V, et al. Investigations on carboxy dibenzylidene sorbitol hydrogels using EPR spectroscopy. Applied Magnetic Resonance. 2015;**46**:1395-1407. DOI: 10.1007/s00723-015-0690-3

[86] Hinderberger D. EPR spectroscopy in polymer science. Topics in Current Chemistry. 2011;**321**:67-89. DOI: 10.1007/128_2011_236

[87] Kurzbach D, Schömer M, Wilms VS, Frey H, Hinderberger D. How structure-related collapse mechanisms determine nanoscale inhomogeneities in thermoresponsive polymers. Macromolecules. 2012;**45**:7535-7548. DOI: 10.1021/ma3014299

[88] Mocanu S, Matei I, Ionescu S, Tecuceanu V, Marinesc G, Ionita P, et al. Complexation of β-cyclodextrin with dual molecular probes bearing fluorescent and paramagnetic moieties linked by short polyether chains. Physical Chemistry Chemical Physics. 2017;**19**:27839-27847. DOI: 10.1039/C7CP05276F

[89] Mocanu S, Matei I, Leonties A, Tecuceanu V, Hanganu A, Minea Z, et al. New flexible molecular probes bearing dansyl and TEMPO moieties for host–guest interactions in solution and gels. New Journal of Chemistry. 2019;**43**:11233-11240. DOI: 10.1039/C9NJ01554J

[90] Dodero A, Schlatter G, Hébraud A, Vicini S, Castellano M. Polymer-free cyclodextrin and natural polymer-cyclodextrin electrospun nanofibers: A comprehensive review on current applications and future perspectives. Carbohydrate Polymers. 2021;**264**:118042. DOI: 10.1016/j.carbpol.2021.118042

[91] Liu Y, Lin T, Cheng C, Wang Q, Lin S, Liu C, et al. Research progress on synthesis and application of cyclodextrin polymers. Molecules. 2021;**26**:1090. DOI: 10.3390/molecules26041090

[92] Yao X, Huang P, Nie Z. Cyclodextrin-based polymer materials: From controlled synthesis to applications. Progress in Polymer Science. 2019;**93**:1-35. DOI: 10.1016/j.progpolymsci.2019.03.004

[93] Petitjean M, García-Zubiri IX, Isasi JR. History of cyclodextrin-based polymers in food and pharmacy: A review. Environmental Chemistry Letters. 2021;**19**:3465-3476. DOI: 10.1007/s10311-021-01244-5

[94] Jansook P, Loftsson T. Self-assembled γ-cyclodextrin as nanocarriers for enhanced ocular drug bioavailability. International Journal of Pharmaceutics. 2022;**618**:121654. DOI: 10.1016/j.ijpharm.2022.121654

[95] Tian B, Liu J. The classification and application of cyclodextrin polymers: A review. New Journal of Chemistry. 2020;**44**:9137-9148. DOI: 10.1039/C9NJ05844C

[96] Wankar J, Kotla NG, Gera S, Rasala S, Pandit A, Rochev YA. Recent advances in host–guest self-assembled cyclodextrin carriers: Implications for responsive drug delivery and biomedical engineering. Advanced Functional Materials. 2020;**30**:1909049. DOI: 10.1002/adfm.201909049

[97] Jin S, Huang J, Chen X, Gu H, Li D, Zhang A, et al. Nitric oxide-generating antiplatelet polyurethane surfaces with multiple additional biofunctions via cyclodextrin-based host–guest interactions. ACS Applied Bio Materials. 2020;**3**:570-576. DOI: 10.1021/acsabm.9b00969

[98] Matencio A, Navarro-Orcajada S, García-Carmona F, López-Nicolás JM. Applications of cyclodextrins in food science. A review. Trends Food Science and Technology. 2020;**104**:132-143

[99] Velázquez-Contreras F, Zamora-Ledezma C, López-González I, Meseguer-Olmo L, Núñez-Delicado E, Gabaldón JA. Cyclodextrins in polymer-based active food packaging: A fresh look at nontoxic, biodegradable, and sustainable technology trends. Polymers (Basel). 2021;**14**:104. DOI: 10.3390/polym14010104

[100] Lamping S, Stricker L, Ravoo BJ. Responsive surface adhesion based on host–guest interaction of polymer brushes with cyclodextrins and arylazopyrazoles. Polymer Chemistry. 2019;**10**:683-690. DOI: 10.1039/C8PY01496E

[101] Chen X, Taguchi T. Enhanced skin adhesive property of α-cyclodextrin/nonanyl group-modified poly(vinyl alcohol) inclusion complex film. Carbohydrate Polymers. 2021;**263**:117993. DOI: 10.1016/j.carbpol.2021.117993

[102] Sanbhal N, Mao Y, Sun G, Xu RF, Zhang Q, Wang L. Surface modification of polypropylene mesh devices with cyclodextrin via cold plasma for hernia repair: Characterization and antibacterial properties. Applied Surface Science. 2018;**439**:749-759. DOI: 10.1016/j.apsusc.2017.12.192

[103] Sun X, Du Y, Zhao S, Huang Z, Feng Z. Enantioseparation of propranolol, amlodipine and metoprolol by electrochromatography using an open tubular capillary modified with β-cyclodextrin and poly(glycidyl methacrylate) nanoparticles. Microchimica Acta. 2019;**2019**:128-186. DOI: 10.1007/s00604-018-3163-1

[104] Casulli MA, Taurino I, Carrara S, Hayashita T. Integration methods of cyclodextrins on gold and carbon electrodes for electrochemical sensors. C-J Carbon Research. 2019;**5**:78-93. DOI: 10.3390/c5040078

[105] Zilberg RA, Maistrenko VN, Kabirova LR, Dubrovsky DI. Selective voltammetric sensors based on composites of chitosan polyelectrolyte complexes with cyclodextrins for the recognition and determination of atenolol enantiomers. Analytical Methods. 2018;**10**:1886-1894. DOI: 10.1039/C8AY00403J

[106] Utzeri G, Matias PMC, Murtinho D, Valente AJM. Cyclodextrin-based nanosponges: Overview and opportunities. Frontiers in Chemistry. 2022;**10**:859406. DOI: 10.3389/fchem.2022.859406

[107] Tian B, Hua S, Tian Y, Liu J. Cyclodextrin-based adsorbents for the removal of pollutants from wastewater: A review. Environmental Science and Pollution Research. 2021;**28**(2):1317-1340. DOI: 10.1007/s11356-020-11168-2

[108] Wang Z, Chen H, Gao X, Hu B, Meng Q, Zhao C, et al. A novel self-floating cyclodextrin-modified polymer for cationic dye removal: Preparation, adsorption behavior and mechanism. Separation and Purification Technology. 2022;**290**:120838. DOI: 10.1016/j.seppur.2022.120838

[109] Li H, Qi S, Li X, Qian Z, Chen W, Qin S. Tetrafluoroterephthalonitrile-crosslinked β-cyclodextrin polymer as a binding agent of diffusive gradients in thin-films for sampling endocrine disrupting chemicals in water. Chemosphere. 2021;**280**:130774. DOI: 10.1016/j.chemosphere.2021.130774

[110] Qin X, Bai L, Tan Y, Li L, Song F, Wang Y. β-Cyclodextrin-crosslinked polymeric adsorbent for simultaneous removal and stepwise recovery of organic dyes and heavy metal ions: Fabrication, performance and mechanisms. Chemical Engineering Journal. 2019;**372**:1007-1018. DOI: 10.1016/j.cej.2019.05.006

[111] Wang J, Yang F. Preparation of 2-hydroxypropyl-β-cyclodextrin polymers crosslinked by poly(acrylic acid) for efficient removal of ibuprofen. Materials Letters. 2021;**284**:128882. DOI: 10.1016/j.matlet.2020.128882

[112] Yadav S, Asthana A, Singh AK, Chakraborty R, Sree Vidya S, Susan MABH, et al. Adsorption of cationic dyes, drugs and metal from aqueous solutions using a polymer composite of magnetic/β-cyclodextrin/activated charcoal/Na alginate: Isotherm, kinetics and regeneration studies. Journal of Hazardous Materials. 2021;**409**:124840. DOI: 10.1016/j.jhazmat.2020.124840

[113] Wang S, Liu J, Zhao H, Zhang F. Carboxymethyl chitosan crosslinked β-cyclodextrin containing hydrogen bonded N C QDs nanocomposites to design fluorescence probes for manganese ion (II) sensing. Materials Science and Engineering: C. 2021;**119**:111556. DOI: 10.1016/j.msec.2020.111556

[114] Soltani S, Emadi R, Javanmard SH, Kharaziha M, Rahmati A. Shear-thinning and self-healing nanohybrid alginate-graphene oxide hydrogel based on guest-host assembly. International Journal of Biological Macromolecules. 2021;**180**:311-323. DOI: 10.1016/j.ijbiomac.2021.03.086

[115] Sun J, Chen J, Bi Y, Xiao Y, Ding L, Bai W. Fabrication and characterization of β-cyclodextrin-epichlorohydrin grafted carboxymethyl chitosan for improving the stability of Cyanidin-3-glucoside. Food Chemistry. 2022;**370**:130933. DOI: 10.1016/j.foodchem.2021.130933

[116] Sagith P, Reshmi CR, Sundaran SP, Binoy A, Mishra N, Sujth A. β-Cyclodextrin functionalized polyurethane nano

fibrous membranes for drug delivery. Journal of Drug Delivery Science and Technology. 2021;**65**:102759. DOI: 10.1016/j.jddst.2021.102759

[117] Zhao J, Ma M, Yan X, Wan D, Zeng Z, Yu P, et al. Immobilization of lipase on β-cyclodextrin grafted and aminopropyl-functionalized chitosan/Fe3O4 magnetic nanocomposites: An innovative approach to fruity flavor esters esterification. Food Chemistry. 2022;**366**:130616. DOI: 10.1016/j.foodchem.2021.130616

[118] Srivastava B, Singh H, Khatri M, Singh G, Arya SK. Immobilization of keratinase on chitosan grafted-β-cyclodextrin for the improvement of the enzyme properties and application of free keratinase in the textile industry. International Journal of Biological Macromolecules. 2020;**165**:1099-1110. DOI: 10.1016/j.ijbiomac.2020.10.009